Anonymous

Transactions of the Scientific Association

Vol. 13

Anonymous

Transactions of the Scientific Association
Vol. 13

ISBN/EAN: 9783337415839

Printed in Europe, USA, Canada, Australia, Japan

Cover: Foto ©berggeist007 / pixelio.de

More available books at **www.hansebooks.com**

TRANSACTIONS

OF THE

SCIENTIFIC ✳ ASSOCIATION,

❃MERIDEN, CONN.❃

1884.

VOL. 1.

MERIDEN, CONN,
FRANCIS ATWATER, PRINTER.
1885.

CONTENTS.

OFFICERS.

REV. J. H. CHAPIN, Ph. D. - - - PRESIDENT.

REV. J. T. PETTEE, A. M. - - - VICE-PRESIDENT.

CHAS. H. S. DAVIS, M. D - - - SECRETARY.

ALBERT B. MATHER, A. M. - - TREASURER.

ROBERT BOWMAN, - CURATOR AND LIBRARIAN.

DIRECTORS OF SECTIONS

Geology and Palæontology, - - - Rev. J. H. Chapin.

Astronomy, - - - - - Rev. J. T. Pettee.

Archæology and Ethnology, - - - Dr. C. H. S. Davis.

Anthropology, - - - - - M. A. Stone, A. M.

Biology, - - - - - Rev. A. H. Hall.

Microscopy, - - - - - - F. J. Seidensticker.

Botany, - - - - - - Mrs. H. H. Kendrick.

Zoology, - - - - - A. B. Mather, A. M.

Geography, - - - - - Miss Ida J. Hall.

Chemistry, - - C. E. Skidgell and Miss Kate R. Kelsey.

Mechanics, - - - - - Henry S. Pratt, A. M.

Technology, - - - - - - Rev. J. V. Garton.

Electrical Science, - - - - Waldo L. Upson.

Entomology, - - - - - - Henry Hirons.

Ichthyology. - - - - - - E. B. Everitt.

Conchology, - - - - - - Miss Ella Daniels.

Necrologist, - - - - - Rev. J. S. Breckenridge.

PUBLISHING COMMITTEE.

Mrs. H. H. Kendrick,   Chas. H. S. Davis, M. D.

Miss Kate R. Kelsey.

REPORT OF SECRETARY.

The Association has now completed its fourth year. Beginning in 1880 with eleven members it now has a membership of one hundred and seventeen.

During the past year five gentlemen and five ladies were elected to membership, two members have resigned and one member has died.

The directors of the various sections have performed their work faithfully, and have presented at each alternate month a condensed report of the progress of investigation and research made during the previous month.

The public meetings have usually been well attended. During 1884, the following papers were read at the public meetings of the association:

FEBRUARY

Sound,	Dr. G. H. Wilson.
Stenography,	J. H. Mabbett.
Geological Ages,	Rev. J. H. Chapin.

APRIL.

Science on the Farm,	Prof. L. P. Chamberlain
Chemical Affinities,	Chas. E. Skidgell.
Red Sunsets,	Rev. J. T. Pettee.

JUNE.

Spectrum Analysis,	Prof. H. S. Pratt.
Myths,	Miss Emily J. Leonard.
Ghosts and Other Illusions and Delusions,	Dr. C. H. S. Davis.

OCTOBER.

Evolution,	Rev. J. S. Breckenridge.
Newspaper Printing,	J. H. Mabbett.
A Plant Parasite,	F. J. Seidensticker.

DECEMBER.

Lepidoptera,	Henry Hirons.
Evolution in the Light of the New Church,	E. B. Everitt.
Description of Recent Geological Purchases,	Rev. J. H. Chapin.

During the summer an excursion was made to Little Falls, near Durham, and some very fine specimens of ichthyolites were obtained.

The September meeting of the Association was devoted to the memory of Miss Emily J. Leonard, who died in July. A memorial address was delivered by Dr. Chas. H. S. Davis, and papers on the life and character of Miss Leonard were read by Rev. J. T. Pettee, Mrs. J. T. Pettee, Mrs. H. H. Kendrick, and Mr. J. H. Mabbett, and remarks were made by Rev. J. H. Chapin.

The Association may well feel satisfied with its work during the past year, and it has established the fact that it is an educational force in the community, and represents, to a large extent, the intellectual life of this city. But there is yet need of developing a more thorough and practical interest in the work of the Association among the citizens.

It is hoped that the museum of the Association will be utilized by the pupils in our public schools, and by others interested in natural science.

CHAS. H. S. DAVIS,
Secretary.

REPORT OF TREASURER.

RECEIPTS.

Balance from 1883,	.	$278 48
Received from dues,	.	33 00
		$311 48

EXPENDITURES.

For apparatus, specimens, etc..		$120 50
Current expenses,	.	38 00
Balance on hand,	.	152 98
		$311 48

A. B. MATHER,
Treasurer.

REPORT OF CURATOR AND LIBRARIAN.

I have to report the following additions to our Association during the year 1884:

One specimen of silver ore from St. Nicholas mine, Mexico.

Three specimens of copper ore.

Specimens of fossils, chiefly Brachiopoda, probably of Silurian age.

Specimen fossil leaf and shells, from Platt Cañon and Greeley, Colorado.

Specimen natural brick and porcelain or baked clay from Bad Lands, Dakota.

Specimen of actinolites.

Specimen of Schistose rock.

Some 150 specimens from line of Union Pacific railroad, purchased for the Association by Rev. J. H. Chapin.

Specimens of fossil fish from Middlefield, Conn.

Four volumes of Astronomical papers of The American Ephemeris.

Three volumes United States Geological Surveys.

One volume Geology of the Comstock Lode.

Transactions of Vassar Brothers' Institute, 1883–1884, Vol. 2.

Annual Report of the Board of Regents of the Smithsonian Institution, 1882.

Two volumes Government Atlas of the Comstock Lode and Washoe District.

ROBERT BOWMAN,
Curator and Librarian.

PROCEEDINGS. 1884.

Regular meeting of the Association, January 14, 1884. Mr. and Mrs. George M. Clark were elected members of the Association. Miss Mary Moses and Mr. Elmer Stannard were proposed for membership.

Voted, That the Treasurer prepare and have printed a form of notification to members in relation to unpaid dues.

Voted, That the Treasurer be instructed to procure a book for permanent records and to transfer the past records.

The following officers were then elected for the ensuing year:

> *President*—Rev. J. H. CHAPIN.
> *Vice-President*—Rev. J. T. PETTEE.
> *Secretary*—Dr. C. H. S. DAVIS.
> *Treasurer*—A. B. MATHER.
> *Curator and Librarian*—ROBERT BOWMAN.

The committee on programme for the February meeting was continued, after reporting progress.

Voted, That the appointment of directors of sections be left to the three first officers of the Asssociation.

Adjourned to meet Monday evening, February 11, 1884.

Regular meeting of the Association, February 11, 1884. Miss Mary Moses and Mr. Elmer Stannard were elected members of the Association.

Mr. Lawrence B. Stevens was proposed for membership.

The report of the lecture committee was accepted and the committee discharged.

Messrs. Pettee, Barber and Miss Leonard were appointed a committee to collect information for the Ornithologists' Union.

Messrs. Pettee, Quested, and Miss Kelsey were appointed a committee on programme for the April meeting.

Adjourned to meet Monday evening, March 10, 1884.

———

March 10, 1884.

Regular meeting of the Association, Monday evening, March 10, 1884.

Lawrence B. Stevens was elected a member of the Association.

Committee on programme reported, and committee continued.

Voted, That the Association purchase the jet for the stereopticon procured by Dr. Chapin.

Adjourned to meet Monday evening, April 14, 1884.

———

April 14, 1884.

Regular meeting of the Association, Monday evening, April 14, 1884.

Voted, That Mrs. Kendrick be appointed one of the directors of the botanical section.

Voted, That Miss Leonard and Mrs. Kendrick be appointed a committee to procure books for preserving botanical specimens.

Miss Emma A. Warner was proposed for membership.

Adjourned to meet Monday evening, May 12, 1884.

———

May 12, 1884.

Regular meeting of the Association, Monday evening, May 12, 1884.

Miss Emma A. Warner was elected a member of the Association.

Mr. H H. Kendrick was proposed for membership.

Committee on programme reported, and committee continued.

Voted, That the Secretary call for the reports of the directors of sections in their reverse order from the previous meeting.

Messrs. Chapin, Davis and Bowman were appointed a committee on excursion.

Adjourned to meet Monday evening, June 9, 1884.

———

June 9, 1884.

Regular meeting of the Association, June 9, 1884.

Mr. H. H. Kendrick was elected a member of the Association.

Rev. J. H. Breckenridge was proposed for membership.

The report of the committee on programme was accepted and the committee discharged.

Messrs. Garton, Catlin, and Miss Daniels were appointed committee on programme for the October meeting.

Adjourned to meet Monday evening, September 8, 1884.

———

JULY 21, 1884.

At a special meeting of the Association held at the residence of the President, Rev. J. H. Chapin, on the 21st day of July, 1884, for the purpose of taking some action in regard to the adoption of appropriate resolutions in relation to the death of their late associate, Miss Emily J. Leonard, the following members were present: Mrs. Kendrick, Mrs. Griswold, Mrs. Hayes, Miss Hattie Linsley, Messrs. Chapin, Pettee, Griswold, F. J. and F. R. Seidensticker, Hayes, Robinson and Bowman. The meeting was called to order by the President, who briefly stated the object of the meeting.

In the absence of the Secretary, Robert Bowman was appointed pro tem.

On motion of Mr. F. J. Seidensticker, a committee was appointed by the chair to draft suitable resolutions. Said committee consisted of Messrs. Pettee and F. J. Seidensticker, and Mrs. Kendrick, who presented the following resolutions, which were unanimously adopted:

WHEREAS, The botanist of this Association, Miss Emily J. Leonard, has been laid to rest beneath the wild flowers which she loved so well; therefore,

Resolved, That we recognize in Miss Leonard a woman of rare intellectual endowments, a fine classical scholar, a most devoted and enthusiastic student of nature, and a most faithful and laborious member of this Association.

Resolved, That in her favorite field of botanical research, in town or state, Miss Leonard had few superiors. Our own herbarium attests the accuracy of her botanical knowledge no less than her untiring industry, while the recognition of her personal discoveries by eminent botanists of this and other states, shows the appreciation of her collections abroad. And, now, that her last collection has been made, our fields and mountains will miss her familiar footsteps, our wild flowers will weep, and Flora herself mourn for her loved disciple.

Resolved, That as members of the Meriden Scientific Association, to which Miss Leonard was so devoted, and to whose exercises her carefully prepared papers contributed so much of knowledge and interest, if we may not hope to equal her in the richness and originality of contributions, we will endeavor to imitate her in her zeal and devotion to truth, and in the promptitude with which she responded to all the calls of the Association.

Resolved, That at the next meeting of the Association, Monday evening, Sept 8th, Dr. C. H. S. Davis be requested to present a sketch of the life and scientific labors of Miss Leonard; that other members be invited to speak *in memoriam,* and that the meeting itself be memorial with reference to her life and death.

SEPTEMBER 8, 1884.

Regular meeting of the Association, Monday evening, September 8, 1884.

Rev. J. H. Breckenridge was elected a member of the Association.
Mrs. F. H. Dilloway was proposed for membership.

Voted, That the Association make an excursion to Little Falls, and that Messrs. Seidensticker and Bowman be appointed a committee to arrange for the excursion.

Adjourned to meet Monday evening, October 13, 1884.

———

NOVEMBER 10. 1884.

Regular meeting of the Association, Monday evening, November 10, 1884.

Committee on programme for the December meeting reported, and committee continued.

Voted, That the curator procure new shelves for the cabinet.

Adjourned to meet Monday evening, Dec. 8, 1884.

———

DECEMBER 8, 1884.

Regular meeting of the Association, Monday evening, December 8, 1884.

Miss Kate Burroughs was elected a member of the Association.

The following persons were proposed for membership: Eli C. Birdsey, Miss Hettie Lewis, Mr. and Mrs. L. L. Sawyer, Miss Sarah N. Benedict, Rev. Charles H. Everest, Mr. and Mrs. H. B. Allen, Jas. F. Allen, Miss Cara G. Allen.

Voted, That the committee on programme be discharged.

Voted, That Mr. Hirons be appointed director of the department of entomology.

Voted, That the Treasurer be authorized to pay Dr. Chapin's bill $99.50, for the purchase of materials.

Messrs. Stone, Quested, and Miss Hattie Bradley were appointed committee on programme for the February meeting.

Adjourned to meet Monday evening, January 12, 1885.

Emily Josephine Leonard was born in Meriden, August 21, 1837, and died there July 16, 1884, in the 47th year of her age. The greater part of her life after 1853, was spent in teaching in public and private schools. She passed the Harvard examination for women, and was offered a professorship at Vassar College.

Miss Leonard was well read in Greek and Latin, also spoke French and German, and had a sufficient knowledge of Italian and Spanish to read in those languages. She assisted Professor Boches, of Harvard University, in the preparation of his French grammar, and translated from the French a History of Political Economy in Europe, by Jerome Adolphe Blanqui. Miss Leonard also assisted Mr. John J. Lalve, in the preparation of his Cyclopedia of Political Economy.

Miss Leonard was greatly interested in the subject of political economy, and accumulated a valuable library on the subject. An essay on "Money," read by her at a convention in Cleveland, Ohio, attracted a great deal of attention at the time, as did also a paper on "Political Economy," read at Portland, Maine, before the Woman's Congress in 1882.

In 1880, when the Scientific Association was formed, Miss Leonard was one of the earliest and most active members. One of her first papers read before the Association was on the "Definition of Botanical Terms." In February, 1880, she read a paper on "Pollen, and the Means by Which It is Distributed." In 1882, at the Darwin meeting of the Association, she read a paper on "Darwin's Observations and Experiments."

December 11, 1882, she read a paper on "Stomata and their Functions." June 11, 1883, she read a paper on "Dimorphous and Trimorphous Heterostyled Plants." She read at other meetings of the Association, papers on the "Nutrition of Plants," "Fertilization of Plants," and in June, 1884, a paper on "Myths and Myth Makers."

The catalogue of plants contained in this volume was prepared with her usual thoroughness, and contains seven hundred and forty-nine distinct species of plants found growing in Meriden. Miss Leonard was both earnest and painstaking in her work in this department of science, and was animated by a strong desire to add something to our knowledge of our native flora.

CATALOGUE

OF THE

PHÆNOGAMOUS AND VASCULAR

CRYPTOGAMOUS PLANTS,

FOUND GROWING IN MERIDEN, CONN.

(INCOMPLETED)

BY

EMILY J. LEONARD,

(DIRECTOR OF THE BOTANICAL DEPARTMENT OF THE MERIDEN
SCIENTIFIC ASSOCIATION.)

—

Published by the Meriden Scientific Association.

—

MERIDEN, CONN,
FRANCIS ATWATER, PRINTER.
1885.

SUMMARY OF SPECIES.

Polypetalous,	257	
Monopetalous,	249	
Apetalous,	82—588	
Gymnosperms,	4	
Exogens,	592	
Endogens,	121	
Phænogamous Plants,		713
Cryptogamous,		36
Total,		749

PREFACE.

---•◦•---

This Catalogue, compiled by Miss EMILY J. LEONARD, is published by the Meriden Scientific Association as a tribute to her worth as a member of the society, and also as a recognition of her value as a botanist.

It is published as she left it, and notwithstanding its incompleteness, will doubtless prove valuable to those who will continue the work she has begun.

To the public, who have little knowledge, and less appreciation of a work of this kind, we would say that it represents at least five years of the constant labor of one woman, almost wholly unaided.

Miss LEONARD's death, at the beginning of the summer's work which was to have completed the catalogue, is a sufficient apology for any errors it may possibly contain. There are some plants catalogued which are not found in Gray, Torrey, or Wood, but as the compiler had access to other botanies, they are probably found in works with which we are not familiar.

The following plants were found in Meriden by Miss LEONARD, and are not mentioned in the "Catalogue of Flowering Plants growing within thirty miles of Yale College":—

Helianthemum corymbosum.
Stellaria pubera.
Vitis vulpina, v. indivisa, v. bipinnata.
Baptisia australis.
Spiræa lobata.

Agrimonia parviflora.

Lonicera grata.

Lonicera ciliata.

Coreopsis lanceolata.

Azalea arborescens.

Scutellaria saxatilis.

Lilium Catesbæi.

We, knowing the thoroughness of her methods, have no hesitancy in submitting this work to any fair criticism.

THE EDITORS.

MARCH 18, 1885.

SERIES I.
PHÆNOGAMOUS OR FLOWERING PLANTS.

———

Class I.   DICOTYLEDONOUS or EXOGENOUS PLANTS.

———

Subclass I.   ANGIOSPERMÆ.

Division I.   POLYPETALOUS EXOGENOUS PLANTS.

———

Order I.   RANUNCULACEÆ.

1. CLEMATIS.

verticillaris. Woods near the Poor-house. 1882
Virginiana. Common.

2. ANEMONE.

Virginiana. Common.
nemorosa. Common.

3. HEPATICA.

triloba. Common.   On some plants the late leaves were acutely lobed.

4. THALICTRUM.

anemonoides. Common.
purpurascens. Common.
dioicum. Common.
Cornuti. Common.
clavatum.

5. RANUNCULUS.

abortivus. Common.
recurvatus. Common.

BULBOSUS. Common.

ACRIS. Common.

6. CALTHA.

palustris. Common.

7. COPTIS.

trifolia. Somewhat rare. Found in woods near the Glass Works.

8. AQUILEGIA.

Canadensis. East of Black Pond. Also, near Poor-house, etc.

9. DELPHINIUM.

CONSOLIDA. Sparingly escaped from gardens.

10. ACTÆA.

alba. Not rare.

rubra. Near reservoir.

11. CIMICIFUGA.

racemosa. Not common.

Order 2. MAGNOLIACEÆ.

1. LIRIODENDRON.

Tulipifera.

Order 3. MENISPERMACEÆ.

1. MENISPERMUM.

Canadense. Road up West Peak. Also. East of Black Pond.

Order 4. BERBERIDACEÆ.

1. BERBERIS.

VULGARIS Sparingly found on the mountains.

2. CAULOPHYLLUM.

thalictroides. South Meriden. May 16. 1883.

Order 5. NYMPHÆACEÆ.

1. NYMPHÆA.

odorata.

2. NUPHAR.

advena. Common.

Order 6. SARRACENIACEÆ.

1. SARRACENIA.

purpurea. South of Black Pond. In bloom June 3, 1883.

Order 7. PAPAVERACEÆ.

1. ARGEMONE.

Mexicana. Sparingly escaped from gardens.

2. CHELIDONIUM.

majus.

3. SANGUINARIA.

Canadensis.

Order 8. FUMARIACEÆ.

1. DICENTRA.

Cucullaria. Cathole Pass. Northwest of Park. East Mountain, etc.

2. CORYDALIS.

glauca. On rocks Southeast of Black Pond. Also, on West Peak.

3. ADLUMIA.

cirrhosa. Cathole Pass.

4. FUMARIA.

officinalis. Escaped from old gardens.

Order 9. CRUCIFERÆ.

1. NASTURTIUM.

officinale.

palustre.

Armoracia. Ditch opposite Mr. S. Cone's, East Meriden.

2. DENTARIA.

diphylla. East of Black Pond, Spruce Glen, and elsewhere. Common.

3. CARDAMINE.

rhomboidea. West of Trotting Park. Also, near Bradley & Hubbard's.

4. ARABIS.

lyrata. Rocks Southeast of Black Pond. Also, on West Peak. June, 1883.
Canadensis. Mountain side, path to West Peak. June 21, 1883.
lævigata. Mountain side, path to West Peak. June 21, 1883.

5. BARBAREA.

vulgaris. Stem leaves sagittate clasping. May 28, 1882. South Meriden.

6. ERYSIMUM.

cheiranthoides. Found in an overgrown path on the Lawrence Place, Aug. 8, 1883.

7. SISYMBRIUM.

OFFICINALE. Common.

8. BRASSICA.

oleracea. Tall form with small leaves. Open field on Pleasant street. May, 1883.
SINAPISTRUM.
NIGRA.
CAMPESTRIS.

9. CAPSELLA.

BURSA-PASTORIS. Common.

10. LEPIDIUM.

Virginicum. Very common
intermedium. 1881.

11. RAPHANUS.

RAPHANISTRUM.
SATIVUS.

Order 10. VIOLACEÆ.

1. VIOLA.

lanceolata. Rare.
blanda. Common.
cucullata. Common.

var. **palmata.**

var. **cordata.**

sagittata.

var. **ovata.**

pedata.

rostrata. Spur ½ inch long, measured from beneath. Woods Northwest of Trotting Park. May 10, 1883.

canina. Common.

Muhlenbergii, Torr.

pubescens.

TRICOLOR. Escaped from gardens.

Order 11. CISTACEÆ.

1. LECHEA.

major.

minor, var. **gracilis.** East Meriden. September 2, 1882.

Novæ-Cæsareæ.

2. HELIANTHEMUM.

corymbosum.

Canadense.

Order 12. DROSERACEÆ.

1. DROSERA.

rotundifolia. Mrs. N. M. Leonard's, East Meriden. July 30, 1882.
(To this plant Darwin devotes 277 pages of his work on Insectivorous Plants.)

Order 13. HYPERICACEÆ.

1. HYPERICUM.

PERFORATUM. Common.

corymbosum. Common.

mutilum. Rather common in low grounds. Near Black Pond.

Canadense.

Sarothra. Open field North of Mr. James Breckenridge's, on Pleasant street.

2. ELODES.

Virginica. October 4, 1881, in fruit.

Order 14. CARYOPHYLLACEÆ.

1. DIANTHUS.

ARMERIA.

2. SAPONARIA.

OFFICINALIS. Common by roadsides near old places.
inflata.*

3. SILENE.

antirrhina.
NOCTIFLORA.

4. LYCHNIS.

GITHAGO.

5. STELLARIA.

MEDIA. Common.
longifolia.
pubera.

6. HOLOSTEUM.

UMBELLATUM.

7. CERASTIUM.

VULGATUM. Common.

8. SPERGULARIA.

rubra.

9. SPERGULA.

ARVENSIS.

10. SCLERANTHUS.

ANNUUS. Roadsides. September 16, 1881.

11. MOLLUGO.

verticillata. Common near sidewalks.

Order 15. PORTULACACEÆ.

1. PORTULACA.

OLERACEA. Very common.

---

*Not catalogued by Gray, Torrey, or Wood.—EDITORS.

2. CLAYTONIA.

Virginica.
Caroliniana.

Order 16. MALVACEÆ.

1. MALVA.

ROTUNDIFOLIA.
MOSCHATA. Escaped from gardens.

2. HIBISCUS.

TRIONUM. Escaped from gardens.

Order 17. TILIACEÆ.

1. TILIA.

Americana.

Order 18. LINACEÆ.

1. LINUM.

USITATISSIMUM.

Order 19. GERANIACEÆ.

1. GERANIUM.

maculatum. Common.
    var. **alba.*** Single plant found in East Meriden in 1881. Reported also, as found in 1880.
Robertianum. Found on the mountains, near the base.

2. IMPATIENS.

fulva. Common by brooksides.
pallida.

3. OXALIS.

stricta. Common.

Order 20. RUTACEÆ.

1. ZANTHOXYLUM.

Americanum. Rather rare.

---

*Not catalogued by Gray, Torrey, or Wood. Probably named by Miss Leonard.—EDITORS.

Order 21. ANACARDIACEÆ.

1. RHUS.

typhina.
glabra.
copallina.
venenata.
Toxicodendron.

Order 22. VITACEÆ.

1. VITIS.

Labrusca. June 12, 1881.
æstivalis.
cordifolia.
vulpina.
indivisa.
bipinnata.

2. AMPELOPSIS.

quinquefolia. Common.
trifolia.* Common. Rarely more than 2 feet in height.

Order 23. RHAMNACEÆ.

1. CEANOTHUS.

Americanus. Common on the mountains.

Order 24. CELASTRACEÆ.

1. CELASTRUS.

scandens.

Order 25. SAPINDACEÆ.

1. STAPHYLEA.

trifolia. Ascent to West Peak. June 23, 1883.

2. ÆSCULUS.

HIPPOCASTANUM.

---

*Not catalogued by Gray, Torrey, or Wood.—EDITORS.

3. ACER.

Pennsylvanicum.
spicatum.
saccharinum. May 9, 1883. Common.
dasycarpum. April 20, 1883. Common.
rubrum. April 23, 1883. Common.

Order 26. POLYGALACEÆ.

1. POLYGALA.

sanguinea. East Meriden. September 2, 1882.
Nuttallii.
verticillata. August, 1883. September, 1881. Common.
polygama.
paucifolia. South Meriden woods. May 30, 1882.

Order 27. LEGUMINOSÆ.

1. LUPINUS.

perennis.

2. CROTALARIA.

sagittalis.

3. TRIFOLIUM.

ARVENSE. Common.
PRATENSE. Common.
MEDIUM. Near Mr. James Breckenridge's, Pleasant street.
repens. Common.
  var. roseum.* Rare.
AGRARIUM. Common.
PROCUMBENS. Common.

4. MELILOTUS.

OFFICINALIS.
ALBA.

5. ROBINIA.

Pseudacacia.
viscosa. Willow Hill. South side of road, near foot of hill.

---

*Not catalogued by Gray, Torrey, or Wood.—EDITORS.

6. DESMODIUM.

rotundifolium.

Dillenii. In flower and fruit, September 4, 1882. East Meriden.

Canadense. Road to Yalesville. July 31, 1883. Woods South of city. August 13, 1883, 2 feet high; flowers deep crimson. $\frac{1}{4}$–$\frac{1}{2}$ in. long. Racemes, glandular hairy. Leaves with strongly reticulated veins, very hairy beneath.

acuminatum. 1882. Woods South of Pleasant street.

7. LESPEDEZA.

procumbens. East Meriden. August 20, and September 2, 1882.

hirta. East Meriden. August 20, and September 2, 1882.

violacea. var. DIVERGENS.

capitata. Southwest of Old Ladies' Home, South Crown street.

8. APIOS.

tuberosa.

9. PHASEOLUS.

perennis.

10. CLITORIA.

Mariana. Found on the ascent to West Peak.

11. AMPHICARPÆA.

monoica. Common. August 6, 1881

12. BAPTISIA.

tinctoria.

australis. Apparently, though in very early bud. West of the wall West of Trotting Park. May 29, 1883.

13. CASSIA.

Marilandica. Bed of old pond. East Meriden. August 7, 1882

Chamæcrista.

nictitans.

14. GLEDITSCHIA.

triacanthos.

15. SAROTHAMNUS.

scoparius. By roadside, near old John Slain place. Said to have been introduced from Ireland some 20 or more years ago.

Order 28. ROSACEÆ.

1. PRUNUS.

Americana.
SPINOSA.
Pennsylvanica.
Virginiana.
serotina.

2. SPIRÆA.

salicifolia.
tomentosa.
lobata.

3. AGRIMONIA.

Eupatoria.
parviflora.

4. GEUM.

rivale.
album.

5. POTENTILLA.

Norvegica. Common.
Canadensis. Common
argentea. Rather common. West Peak. June 23, 1883.
arguta. Not rare.
tridentata.

6. FRAGARIA.

vesca.
Virginiana.

7. RUBUS.

odoratus. On the mountains.
triflorus. South of old Charles Paddock place. May 23, 1882.
strigosus.
occidentalis. June 1, 1881.
villosus.
Canadensis.
hispidus.
cuneifolius.

8. ROSA.

RUBIGINOSA.
lucida.
MICRANTHA. Rare.
Carolina.

9. CRATÆGUS.

coccinea.
Oxyacantha.
tomentosa.

10. PYRUS.

arbutifolia.
Americana.

11. AMELANCHIER.

✸Canadensis.

Order 29. SAXIFRAGACEÆ.

1. RIBES.

Cynosbati.
hirtellum.
floridum.
aureum.
lacustre.
rubrum. Near the mountain.

2. PARNASSIA.

Caroliniana. Swamp in East part of town. Petals four times the length or sepals.

3. SAXIFRAGA.

Virginiensis. One capsule had 3 carpels; was three-beaked.
Pennsylvanica. A variety found in Kensington had dark pink flowers, and light salmon colored anthers. Very pretty.

4. MITELLA.

diphylla. Cathole Pass. May 21, 1882.

5. TIARELLA.

cordifolia. East Mountain.

6. CHRYSOSPLENIUM.

Americanum. West side of Middletown road, near mountain.

Order 30. CRASSULACEÆ.

1. PENTHORUM.

sedoides. Common.

2. SEDUM.

Telephium.

Order 31. HAMAMELACEÆ.

1. HAMAMELIS.

Virginica.

Order 32. ONAGRACEÆ.

1. CIRCÆA.

Lutetiana.

2. EPILOBIUM.

angustifolium. East of Black Pond. August 6, 1882.
palustre.
coloratum.

Also, a white variety with large ovate-lanceolate leaves.

3. ŒNOTHERA.

biennis. Common. Also, variety muricata.
pumila. Frequent in marshy land.

4. LUDWIGIA.

palustris.
alternifolia. Swamp in East Meriden.

Order 33. MELASTOMACEÆ.

1. RHEXIA.

Virginica. Hairs on ovary and margin to stem, glandular.

Order 34. LYTHRACEÆ.

1. LYTHRUM.

alatum.

Order 35. CUCURBITACEÆ.

1. SICYOS.

angulatus.

Order 36. UMBELLIFERÆ.

1. HYDROCOTYLE.

Americana.

2. SANICULA.

Canadensis.
Marilandica.

3. DAUCUS.

CAROTA.

4. PASTINACA.

SATIVA.

5. ARCHANGELICA.

hirsuta.
atropurpurea.

6. THASPIUM.

aureum.

7. CICUTA.

maculata.

8. SIUM.

lineare. East of Black Pond, where the water had receded in dry weather.
LATIFOLIUM. Near Black Pond.

9. CRYPTOTÆNIA.

Canadensis.

10. OSMORRHIZA.

longistylis.
brevistylis.

11. CONIUM.

MACULATUM.

Order 37. ARALIACEÆ.

I. ARALIA.

racemosa. Southwest part of town.
nudicaulis. Rather common.
trifolia. Common.

Order 38. CORNACEÆ.

1. CORNUS.

Canadensis. Southeast of Black Pond.
florida. Common.
sericea.
stolonifera.
paniculata.
alternifolia.

2. NYSSA.

multiflora.

Division II. MONOPETALOUS EXOGENOUS PLANTS.

Order 39. CAPRIFOLIACEÆ.

1. LONICERA.

sempervirens.
grata.
parviflora.
ciliata. East Meriden. May 4, 1884. Cathole Pass.

2. DIERVILLA.

trifida.

3. TRIOSTEUM.

perfoliatum. Not common.

4. SAMBUCUS.

Canadensis. Common.

5. VIBURNUM.

Lentago.
acerifolium.
nudum.
dentatum.
lantanoides.
prunifolium.

Order 40. RUBIACEÆ.

1. GALIUM.

Aparine.
asprellum.
trifidum.
triflorum.
pilosum.
circæzans.

2. CEPHALANTHUS.

occidentalis.

3. MITCHELLA.

repens.

4. HOUSTONIA.

cærulea.

Order 41. DIPSACEÆ.

1. DIPSACUS.

SYLVESTRIS.

Order 42. COMPOSITÆ.

1. VERNONIA.

Noveboracensis. Common.

2. LIATRIS.

scariosa.

3. EUPATORIUM.

purpureum. Common.
maculatum.
perfoliatum. Common.
ageratoides. Rare. Reddish purple stem and leaf veins. Leaves broad cordate, acuminate.

4. TUSSILAGO.

FARFARA.

5. SERICOCARPUS.

conyzoides.
solidagineus.

6. ASTER.

corymbosus. Stems dark reddish purple. Common.
macrophyllus. Rays 10 or more. Common.
lævis. Rays light purple—not blue.
    var. cyaneus.
undulatus. Common.
cordifolius. Also, var. light sky blue, with margined petioles. October, 1881.
dumosus. September 6, 1882. City.
Tradescanti. September 25, 1882. Common.
simplex. October 7. 1882.
puniceus. September 2, 1882. Also, October 3. East Meriden.
Novæ-Angliæ.
multiflorus.

7. ERIGERON.

Canadense.
annuum.

strigosum.

bellidifolium. Flowers, pale lilac. The first aster to bloom. May 23, 1882.

Philadelphicum.

8. DIPLOPAPPUS.

amygdalinus. Pappus all thickened at the tip, and yellowish white. No
shorter outer pappus discoverable.

cornifolius. Found no shorter outer pappus.

umbellatus.

9. SOLIDAGO.

bicolor. Common ; 7 or 8 rays. August 28, 1881.

latifolia.

cæsia. September 2, 1882.

stricta. August 12, 1882.

sempervirens. August 13, 1882.

patula. September 2, 1882.

arguta. August 20. 1883.

Muhlenbergii. September 2, 1882.

altissima. Common.

var. rugosa. Common.

odora. Reported by Mrs. Davis. Not seen by compiler of catalogue.

nemoralis. September 10, 1882.

Canadensis. var. scabra. August 28, 1881.

serotina. August 13, 1882; August 19, 1883.

lanceolata.

squarrosa.

neglecta.

10. INULA.

HELENIUM. North of Black Pond. Upper leaves somewhat cordate, ovate,
dentate, blunt pointed. Rays 50 or more. August 13, 1882

11. AMBROSIA.

artemisiæfolia. Very common.

trifida. Leaves mostly five-lobed. Veteran street, near Main. Sept. 8, 1882.

12. RUDBECKIA.

hirta.

laciniata.

13. HELIANTHUS.

strumosus. Also, a variety with leaves tapering to the base. Aug. 13, 1882.

divaricatus. Rays hardly an inch long.

decapetalus.

ANNUUS. Sparingly escaped from gardens.

14. COREOPSIS.

lanceolata.

15. BIDENS.

cernua. One foot or more in height. Stem rough, with crooked hairs. Black Pond. October 4, 1881.
frondosa.
chrysanthemoides. Had three or four awns. Black Pond. Oct. 4, 1881.
connata.

16. MARUTA.

COTULA. Common by roadsides.

17. ACHILLEA.

Millefolium. Common.

18. LEUCANTHEMUM.

VULGARE. Common.

19. TANACETUM.

VULGARE. By roadsides.

20. ARTEMISIA.

VULGARIS. Waste places, near dwellings.

21. GNAPHALIUM.

decurrens. Hillsides.
polycephalum. September 3. 1882.
purpureum.
uliginosum. Ditches. Common.

22. ANTENNARIA.

margaritacea. Achenia reddish brown. Half-line or less in length. Common.
plantaginifolia. Common. East of Black Pond. April 30. 1883.

23. SENECIO.

aureus. Common. Southeast part of town. June 4, 1881

24. CENTAUREA.

NIGRA.
CYANUS.
arvense.*

---

*Not catalogued by Gray, Torrey, or Wood.—EDITORS.

25. ERECHTHITES.

hieracifolia. September 2, 1882.

26. CIRSIUM.

LANCEOLATUM.
muticum. October 3.
ARVENSE. July 4, 1881.
> var. **album.*** Near Veteran street. August 20, 1883. Very fragrant. Leaves dark green, glossy, very wavy, spinous. Flowers very small. Heads about one-half to three-fourths in. in diameter.

Virginianum. Found in woods.

27. LAPPA.

OFFICINALIS.
> var. MAJOR. Common by roadsides. Varies in color of flower, from white to pink, dark crimson and variegated. Flowers beautiful.

28. CICHORIUM.

INTYBUS. Roadsides. Rare. July 5, 1882.

29. KRIGIA.

Virginica. West Peak. Scapes hollow. June 21, 1883.

30. LEONTODON.

AUTUMNALE. Meadow on Pleasant street. August 21, 1883.

31. HIERACIUM.

Canadense. September 3, 1882.
scabrum. Rays 19. Disk flowers; 42 on a specimen examined.
venosum. June 28, 1882.

32. NABALUS.

albus. East Meriden. August 29, 1881.
altissimus.

33. TARAXACUM.

Dens-leonis. Common—very.

---

*Not catalogued by Gray, Torrey, or Wood.—EDITORS.

34. LACTUCA.

Canadensis. August 13. 1882.
    var. **integrifolia.** August 13. 1882.
    var. **elongata.** August 13, 1882.
    var. **sanguinea.** August 13, 1882.

35. SONCHUS.

ARVENSIS.
OLERACEUS. August 5, 1882. East Meriden.

Order 43. LOBELIACEÆ.

1. LOBELIA.

cardinalis.
syphilitica.
inflata.
spicata.

Order. 44. CAMPANULACEÆ.

1. CAMPANULA.

Americana.
rotundifolia. Mountain side, Southeast of Black Pond. One specimen had
    26 flowers, and the two double flowers. (October 5, 1881.
    Found, also, August 20, 1883.) The round base leaves had
    disappeared from all but one specimen, leaving none but
    linear leaves.
aparinoides. Wet meadows. 1882.

2. SPECULARIA.

perfoliata.

Order 45. ERICACEÆ.

1. GAYLUSSACIA.

resinosa.
dumosa.

2. VACCINIUM.

vacillans.
corymbosum. var. pallidum. East Meriden. May 25, 1881.
macrocarpon.

3. ARCTOSTAPHYLOS.

Uva-ursi.

4. EPIGÆA.

repens. Spruce Glen.

5. GAULTHERIA.

procumbens.

6. CASSANDRA.

calyculata. Near Black Pond. May 1, 1884.

7. ANDROMEDA.

ligustrina. East Meriden. July 16, 1882.

8. CLETHRA.

alnifolia. Common.

9. KALMIA.

latifolia.
angustifolia.

10. AZALEA.

viscosa.
arborescens.
nudiflora. Common. June 1, 1883.

11. PYROLA.

rotundifolia. Frequent.
elliptica. Rare.

12. CHIMAPHILA.

umbellata.
maculata.

13. MONOTROPA.

uniflora. Not rare. July 16, 1883.
Hypopitys.

Order 46. AQUIFOLIACEÆ.

1. ILEX.

verticillata. Common. July 2, 1883.

2. NEMOPANTHES.

Canadensis. (Found by Mrs. Knapp. Possibly not within the limits of Meriden.)

Order 47. PLANTAGINACEÆ.

1. PLANTAGO.

LANCEOLATA.
MAJOR.

Order 48. PRIMULACEÆ.

1. TRIENTALIS.

Americana. Southeast of Black Pond. June 3, 1883.

2. LYSIMACHIA.

stricta. August 5, 1882. Not common.
quadrifolia. var. with two upper whorls in 4's, others in 2's. Var. in 4's, 5's,
6's and 7's. August 7, 1882. Common.
ciliata. Common. July 23, 1881.
NUMMULARIA. Escaped probably from cultivation.

Order 49. LENTIBULACEÆ.

1. UTRICULARIA.

vulgaris.
clandestina.

Order 50. BIGNONIACEÆ.

1. TECOMA.

radicans.

Order 51. OROBANCHACEÆ.

1. APHYLLON.

uniflorum. June 8, 1883.

Order 52. SCROPHULARIACEÆ.

1. VERBASCUM.

THAPSUS. Common.
BLATTARIA. Rare.

2. LINARIA.

Canadensis.
VULGARIS.

3. SCROPHULARIA.

nodosa. Woods South of Pleasant street August, 1883.

4. CHELONE.

glabra. Common. August, 1883.

5. MIMULUS.

ringens. Common.
alatus.

6. GRATIOLA.

Virginiana.

7. VERONICA.

Anagallis.
officinalis.
serpyllifolia.

8. GERARDIA.

tenuifolia. East Meriden. September 9, 1881.
flava. Road to Yalesville.
quercifolia. Near base of mountain. East shore of Black Pond.
linifolia.* Common.

9. CASTILLEIA.

coccinea. East Meriden.

10. MELAMPYRUM.

Americanum. Common.

11. PEDICULARIS.

Canadensis. Common.
lanceolata.

Order 53. ACANTHACEÆ.

1. DIANTHERA.

Americana. Near Black Pond. October, 1881.

Order 54. VERBENACEÆ.

1. VERBENA.

angustifolia. Yalesville, 1883.
hastata. Common.
urticifolia. Common.

Order 55. LABIATÆ.

1. TRICHOSTEMA.

dichotomum. Frequent.

2. MENTHA.

VIRIDIS. Common.
PIPERITA. Common.
ARVENSIS. Rare. East Meriden. July 16, 1883. In bloom.
Canadensis. Common.
SATIVA. Common.

*Not catalogued by Gray, Torrey, or Wood.—EDITORS.

3. LYCOPUS.

Europæus. Common in marshy land. Near Black Pond. etc.

    var. sinuatus. Lot West side of Pleasant street, South of Mr. Lounsbury's.

4. PYCNANTHEMUM.

incanum.

muticum. Common.

lanceolatum. Northeast part of town. July 27, 1883

linifolium.

5. THYMUS.

SERPHYLLUM. Abundant in the new cemetery, Southwest of city.

6. CALAMINTHA.

Clinopodium. West mountain. June 23, 1883.

7. HEDEOMA.

pulegioides. Frequent.

8. COLLINSONIA.

Canadensis. Common.

9. MONARDA.

didyma. Escaped from cultivation.

fistulosa. Escaped from cultivation.

punctata.

10. NEPETA.

CATARIA. Common.

GLECHOMA. In damp waste places. Probably escaped from cultivation.

11. BRUNELLA.

vulgaris. Common.

12. SCUTELLARIA.

saxatilis.

galericulata. August 13, 1882. Common.

parvula.

lateriflora. South of city. August 13, 1882.

13. LEONURUS.

CARDIACA. Common in waste places.

14. MARRUBIUM.

VULGARE.

15. GALEOPSIS.

TETRAHIT.

16. LAMIUM.

AMPLEXICAULE. In cultivated grounds.
MACULATUM.* In cultivated grounds.

Order 56. BORRAGINACE.E.

1. SYMPHYTUM.

OFFICINALE. Escaped from cultivation. Rare.

2. MYOSOTIS.

palustris.
laxa. Ascent to West Peak. June 23. 1883.

3. CYNOGLOSSUM.

OFFICINALE. Common in East Meriden. June, 1881.
Morisoni. East part of town.

Order 57. HYDROPHYLLACE.E.

1. PHACELIA.

viscida.   A single plant appeared in the garden of Miss H. E. Bradley.
Blossomed the latter part of July. Had the odor of Geranium
Robertianum, and flowers the color of those of Geranium
maculatum, with whitish center.

Order 58. CONVOLVULACE.E.

1. CALYSTEGIA.

sepium.

2. CUSCUTA.

Gronovii. Sprouts in the ground, attaches itself to Jewel weed, etc
tenuiflora.

Order 59. SOLANACE.E.

1. SOLANUM.

DULCAMARA. Thirty-three white seeds in oval berry. Pedicels 3 or 4.
Peduncle 1½ inches.
NIGRUM.

2. LYCIUM.

VULGARE. Escaped from cultivation.

---

*Not catalogued by Gray, Torrey, or Wood.—EDITORS.

3. PHYSALIS.

pubescens. Pratt street. Meriden. September 22, 1881.

4. DATURA.

STRAMONIUM.

Order 60.   GENTIANACEÆ.

1. GENTIANA.

crinita. October 3, 1882.
Andrewsii. Common. October 3, 1881.

Order 61.   APOCYNACEÆ.

1. APOCYNUM.

androsæmifolium.
cannabinum. Field on Cook avenue. June, 1883. Some specimens 6 feet
　　　　in height.

Order 62.   ASCLEPIADACEÆ.

1. ASCLEPAIS.

Cornuti. Common.
phytolaccoides. Woods South of city. June 29. 1883.
purpurascens.
quadrifolia. Thorp's woods. East part of town. 1881.
parviflora. Flowers flesh colored.
variegata.
ovalifolia.
paupercula.
incarnata. var. PULCHRA.
tuberosa.
verticillata. Road to Yalesville. June 30, 1883.

2. PERIPLOCA.

GRÆCA. Found by Miss Derby.

Order 63.   OLEACEÆ.

1. FRAXINUS.

Americana.

Division III.   APETALOUS EXOGENOUS PLANTS.

Order 64.   ARISTOLOCHIACEÆ.

1. ASARUM.

Canadense. Woods South of Black Pond.

2. ARISTOLOCHIA.

Serpentaria. Rare.

Order 65.   PHYTOLACCACEÆ.

1. PHYTOLACCA.

decandra.

Order 66.   CHENOPODIACEÆ.

1. CHENOPODIUM.

HYBRIDUM.
ALBUM.

Order 67.   AMARANTACEÆ.

1. AMARANTUS.

RETROFLEXUS.
ALBUS.

Order 68.   POLYGONACEÆ.

1. POLYGONUM.

Pennsylvanicum.
PERSICARIA. Common.
Hydropiper.
aviculare.
        var. erectum. Very common.
arifolium. Southeast of Black Pond. October, 1881.
sagittatum. In low places.
Virginianum.
cilinode. On rock Southeast of Black Pond. October, 1881.

CONVOLVULUS.
dumetorum. var. **scandens.** August, 1882. Common.

2. FAGOPYRUM.

ESCULENTUM.

3. RUMEX.

CRISPUS.
OBTUSIFOLIUS.
ACETOSELLA.

Order 69.　LAURACEÆ.

1. SASSAFRAS.

officinale.

2. LINDERA.

Benzoin. Common.

Order 70.　SANTALACEÆ.

1. COMANDRA.

umbellata. June 1, 1881. Frequent.

Order 71.　EUPHORBIACEÆ.

1. EUPHORBIA.

maculata. Common.
dentata.
CYPARISSIAS. Escaped from cultivation.
LATHYRIS. Sparingly escaped from gardens.
hypericifolia. Common.

2. ACALYPHA.

Virginica.

Order 72.　URTICACEÆ.

1. ULMUS.

fulva.
Americana.

2. MORUS.

rubra. Rare.
ALBA. Rare. Probably escaped from cultivation.

3. URTICA.

DIOICA. Near Parker's shop, East part of town. September 4, 1882.

4. LAPORTEA.

Canadensis.

5. PILEA.

pumila.

6. PARIETARIA.

Pennsylvanica.

7. HUMULUS.

Lupulus.   Rare.

8. BŒHMERIA.

cylindrica.   East part of town.   July 16, 1883.

Order 73.   PLATANACEÆ.

1. PLATANUS.

occidentalis.

Order 74.   JUGLANDACEÆ.

1. JUGLANS.

cinerea.

2. CARYA.

alba.
porcina.
amara.

Order 75.   CUPULIFERÆ.

1. QUERCUS.

alba.
Prinus.
ilicifolia.   Summit of West Peak.   June 23, 1883.
rubra.
palustris.

2. CASTANEA.

vesca.   Common.

3. FAGUS.

ferruginea.

4. CORYLUS.

rostrata.   Very common.

5. CARPINUS.

Americana. Very common.

6. OSTRYA.

Virginica. Somewhat rare.

Order 76. MYRICACEÆ.

1. MYRICA.

cerifera. June 1, 1881. East part of town.

2. COMPTONIA.

asplenifolia. Common.

Order 77. BETULACEÆ.

1. BETULA.

lenta.
alba, var. populifolia. Leaves doubly serrate. In bloom May 14, 1883.
excelsa, Ait.

2. ALNUS.

serrulata.
incana.

Order 78. SALICACEÆ.

1. SALIX.

tristis. Anthers reddish. Scales black. Middle flowers expanding first.
April, 1883.
Muhlenbergiana, (Torrey). The earliest willow in bloom. Woods south
of Crown street. April, 1883. Staminate flowers open from
apex downward.
sericea. (Two ovaries sessile on a common stipe.). Also a *forma mon-
strosa*. (Marsh.). Stream in woods near Glass Works.
discolor. Fertile catkins 3½ inches long. May 10, 1883. Northwest of
Trotting Park.
PURPUREA.
rostrata. Richardson. livida of Gray.
ALBA. First lvs. entire. Later serrulate. Twigs green. East Meriden.
May 14, 1883.
   var. **vitellina.** (S. vitellina of Torrey.). Willow Hill, East part of
town. May 14 in bloom.
nigra.
lucida. May 24. 1881.
livida. Scales yellowish green, as also stamens, anthers and filaments.
Beautiful.
   var. **occidentalis.** Woods Northwest of Trotting Park. May 10. 1883.

cordata.
FRAGILIS. May 24, 1881.
BABYLONICA. Rare, save in cultivation.

2. POPULUS.

tremuloides.
grandidentata. Common.
balsamifera. East part of town, Middletown road, near Mrs. James Perkins'
house
var. **candicans.** Near houses. Stigma much dilated.

SUBCLASS II. GYMNOSPERMÆ.

Order 79. CONIFERÆ.

1. PINUS.

rigida.
Strobus.

2. ABIES.

Canadensis.

3. TAXUS.

baccata. var. **Canadensis.** Near rocks, Southeast shore of Black Pond. 1882.

CLASS II. MONOCOTYLEDONOUS OR ENDOGENOUS PLANTS.

Order 80. ARACEÆ.

1. ARISÆMA.

triphyllum. Rather common. May 19, 1883.

2. SYMPLOCARPUS.

fœtidus. Common.

3. ACORUS.

Calamus. Common.

Order 81. TYPHACEÆ.

1. TYPHA.

latifolia.

2. SPARGANIUM.

simplex.

Order 82. ALISMACEÆ.

1. ALISMA.

Plantago. var. Americanum. Leaves ovate. Common

2. SAGITTARIA.

variabilis. Common.

Order 83. ORCHIDACEÆ.

1. ORCHIS.

spectabilis. Spur ¼ inch shorter than the ovary. A single spm. from Mrs.
Cone, East part of town May 21. 1883.
flava. (Lindl. Platanthera flava, etc.) 1881.

2. HABENARIA.

orbiculata. Cat Hole, Mrs. Knapp. August 19, 1883. Not in flower.
lacera. Rare.
pyscodes. Rare.
virescens. Rare.
peramœna. Rare.

3. GOODYERA.

pubescens. Frequent.

4. SPIRANTHES.

Romanzoviana.
cernua. Frequent.

5. ARETHUSA.

bulbosa. East part of town. Also in South Meriden. 1881 and 1882. Rare.

6. POGONIA.

ophioglossoides. East Meriden. July 4. 1881. Also 1882 and 1883.
verticillata. Found by Mrs. Knapp. 1883.

7. LIPARIS.

liliifolia. One specimen found on descent from West Peak. June 23. 1883.

8. CORALLORHIZA.

multiflora. August, 1883.

9. CYPRIPEDIUM.

pubescens. Woods near Poor House. Rare.
parriflorum. East part of town, near Mrs. Thorp's. Rare.
acaule.

10. CALOPOGON.

pulchellus. East Meriden, 1881, '82 and '83. July.

Order 84. AMARYLLIDACEÆ.

1. HYPOXYS.

erecta.

Order 85. IRIDACEÆ.

1. IRIS.

versicolor.

2. SISYRINCHIUM.

Bermudiana.

Order 86. SMILACEÆ.

1. SMILAX.

rotundifolia. Common.
glauca.
herbacea. June 12, 1881.

Order 87. LILIACEÆ.

1. TRILLIUM.

cernuum. Northeast part of city. Woods.
erectum. Lowlands near streams.

2. MEDEOLA.

Virginica. East of Black Pond.

3. VERATRUM.

viride.

4. UVULARIA.

perfoliata. Common.
sessilifolia. Common.

5. STREPTOPUS.

amplexifolius. Woods near Hemlock Grove. June, 1883, in seed.

6. SMILACINA.

racemosa.
stellata. Rare.
bifolia. Common. One spm. had the lower leaf briefly petioled, and the upper somewhat clasping.

7. POLYGONATUM.

biflorum. Common.
MULTIFLORUM Occasional.

8. LILIUM.

Philadelphicum.
Canadense.
Catesbæi.

9. ERYTHRONIUM.

Americanum. Common.

10. ORNITHOGALUM.

UMBELLATUM.

11. ALLIUM.

tricoccum. Near Black Pond, East side.
Canadense. Brookside West of Trotting Park. May 29, 1883.

Order 88. MELANTHACEÆ. (Torrey.)

1. HELONIAS.

bullata.

2. CHAMÆLIRIUM.

luteum. Cat Hole Pass. Mrs. Knapp. August 19, 1883.

Order 89. JUNCACEÆ.

1. LUZULA.

campestris. Common. One species (variety?) is cæspitose, about 6 inches in height, has broad lance-linear leaves from 2 to 5 inches long, very hairy, spikes chestnut colored, leaves very purplish at base. Another variety (?) has narrowly linear leaves, is smooth except where the lvs. and bracts join the stem, yellowish spikes and light bluish-green leaves.

2. JUNCUS.

effusus.

tennis. Some capsules were dotted in rows on the margins. Sepals a little longer than the capsule. In fruit, August 4, 1882.

Order 90. PONTEDERIACEÆ.

1. PONTEDERIA.

cordata. Common.

Order 91. ERIOCAULONACEÆ.

1. ERIOCAULON.

septangulare. Black Pond. August 19. 1883.

Order 92. CYPERACEÆ.

1. CYPERUS.

diandrus.
phymatodes.
strigosus.

2. DULICHIUM.

spathaceum.

3. ELEOCHARIS.

tennis. June 20, 1883.
palustris. June 30, 1883.
obtnsa. July.

4. SCIRPUS.

polyphyllus.
Eriophorum.

5. RHYNCHOSPORA.

glomerata.

6. CAREX.

vulpinoidea. (Gray.).
sparganioides.
scoparia. Achenium stalked and crowned with the long continuous style.
intumescens.
debilis.
triceps.
crinita.
stellulata. Nerves of perigynia distinct.
hystricina.
folliculata.
virescens.
retroflexa.
lupulina.
Pennsylvanica. Common in woods. April--July.
digitalis. Scales white at first, tawny later, with green keel and mucronate
point, falling or broken off early.
vestita.
tentaculata. Two capsules at summit of staminate spike. May 29, 1883.
Southwest of park.
stipata. Spms. with anthers *very fragrant* when dry. Marshes. May 9, 1883.

Order 93. GRAMINEÆ.

1. LEERSIA.

oryzoides. Brookside, East Meriden September 4, 1882. Common.

2. PHLEUM.

PRATENSE. Common.

3. AGROSTIS.

vulgaris. Common.
alba.

4. CINNA.

arundinacea.

5. MUHLENBERGIA.

Mexicana. Troublesome in some old gardens. September 3, 1883.
diffusa.

6. DACTYLIS.

GLOMERATA. Common.

7. GLYCERIA.

aquatica. Common.

8. POA.

compressa.
pratensis. Common.
annua.

9. ERAGROSTIS.

PILOSA. Common on Veteran street by sidewalk. August, 1882.
capillaris. East Meriden.

10. FESTUCA.

PRATENSIS. Common. June 14, 1883.
nutans. Common.

11. BROMUS.

SECALINUS.

12. TRITICUM.

repens.
VULGARE.

13. LOLIUM.

PERENNE.

14. DANTHONIA.

spicata.
sericea.

15. ANTHOXANTHUM.

ODORATUM.

16. PHALARIS.

arundinacea. var. picta. Escaped from gardens.
CANARIENSIS. Near sidewalk, Colony street.

17. PANICUM.

SANGUINALE.
proliferum.
capillare. Old witch grass, common in gardens.
latifolium. Common in woods.
pauciflorum.
dichotomum. Woods and low grounds. Common.
depauperatum.
CRUS-GALLI. A purple awned variety, has awns more than an inch long.

18. SETARIA.

GLAUCA.
VIRIDIS.

SERIES II.

CRYPTOGAMOUS ᴏʀ FLOWERLESS PLANTS.

Cʟᴀss III. ACROGENS.

Order 94. EQUISETACEÆ.

1. EQUISETUM.

arvense. Common.
sylvaticum. May 24, 1882. East Meriden.
hyemale. East Meriden. Common.
variegatum.

Order 95. FILICES.

1. POLYPODIUM.

vulgare.

2. ADIANTUM.

pedatum.

3. PTERIS.

aquilina.
    var. **candata.**

4. ALLOSORUS.

atropurpureus.*

5. ASPLENIUM.

Trichomanes. Near rocks and running brook North of town, near railroad.
ebeneum. Same locality as A. Trichomanes.
Filix-fœmina. Common.

6. DICKSONIA.

punctilobula.

---

*Not catalogued by Gray, Torrey, or Wood.—Eᴅɪᴛoʀs.

7. WOODSIA.

obtusa.
Ilvensis.

8. CYSTOPTERIS.

fragalis.

9. ASPIDIUM.

Thelypteris. Common.
Noveboracense. Common.
spinulosum. Common, particularly var intermedium.
Goldianum. Rare. Woods Northeast of city.
marginale. Common.
acrostichoides. Common.

10. ONOCLEA.

sensibilis. Common.

11. LYGODIUM.

palmatum. Quite likely to be found on the mountain, as it is found on the
other side of it, both north and northwest of us.

12. OSMUNDA.

regalis. Common.
Claytoniana. Common.
cinnamomea. Common.

13. BOTRYCHIUM.

lunarioides.
    var. obliquum. (B. ternatum, var. obliquum, Milde.)
    var. dissectum.
Virginicum. (Gray.). Southwest part of town. May 30, 1882.

Order 96. LYCOPODIACEÆ.

1. LYCOPODIUM.

lucidulum.
inundatum.
dendroideum.
clavatum.

2. SELAGINELLA.

apus.

TRANSACTIONS

OF THE

MERIDEN SCIENTIFIC ASSOCIATION·

MERIDEN, CONN.

VOL. II.

1885–1886.

PROCEEDINGS AND TRANSACTIONS

OF THE

SCIENTIFIC ASSOCIATION,

MERIDEN, CONN.

1885–1886.

VOL. II.

MERIDEN, CONN.
E. A. HORTON & CO., PRINTERS.
1887.

CONTENTS.

OFFICERS.

Rev. J. H. Chapin, Ph. D., ... President.

Rev. J. T. Pettee, A. M., ... Vice-President.

Chas. H. S. Davis, M. D., ... Secretary.

Albert B. Mather, A. M..   Treasurer.

Robert Bowman, ... Curator and Librarian.

Mrs. E. B. Kendrick, ... Collector.

DIRECTORS OF SECTIONS.

Geology and Palæontology, ... Rev. J. H. Chapin.

Astronomy, ... Rev. John T. Pettee.

Archæology and Ethnology, ... Dr. Chas. H. S. Davis.

Anthropology, ... Melville A. Stone, A. M.

Biology, ... Miss Wegia H. Hall.

Microscopy, ... Dr. G. H. Wilson.

Botany, ... Mrs. E. B. Kendrick.

Zoology, ... Albert B. Mather, A. M.

Engineering, ... F. E. Sands.

Geography, ... Mrs. Clarence E. Ellsbree.

Chemistry . . . . Chas. E. Skidgell.

Mechanics, ... Henry S. Pratt, A. M.

Technology, ... Rev. J. V. Garton.

Electrical Science, ... Waldo L. Upson.

Entomology, ... Wm. H. Doyle.

Ornithology, ... Franklin Platt.

Ichthyology, ... Edwin B. Everitt.

Conchology, ... Miss M. Ella Daniels.

Necrologist, ... Geo. W. Smith.

PUBLISHING COMMITTEE.

CHAS. H. S. DAVIS., M. D.   FRANKLIN PLATT.

MRS. E. B. KENDRICK.

Report of Secretary.

At the beginning of the sixth year of the Association we have one hundred and thirty members.

During the past year eighteen gentlemen and ten ladies were elected members of the Association.

The Directors of the various sections have prepared with a great deal of care, and presented to the Association every alternate month, a resumé of scientific progress made during the previous two months.

At each alternate month papers have been read by members of the Association and others, to large and appreciative audiences.

The following papers were read before the Association in 1885 :

FEBRUARY.

The Treatment of Horn Silver Ore,	REV. A. H. HALL.
Crystalization, .	PROF. HENRY S. PRATT.

MAY.

Explosives,	MISS KATE R. KELSEY.
Bacteria *vs.* Material Immortality,	DR. G. H. WILSON.

SEPTEMBER.

Our Schools, .	ANDREW J. COE.
The Great Tombs and Smoking Mountains of Mexico,	REV. J. H. CHAPIN.

NOVEMBER.

Review of Pressensé's " History of Origins,"	REV. W. F. MARKWICK.
Review of Drummond's " Natural Law of the Spiritual World,"	REV. A. J. AUBREY.
Review of the Seventh Chapter of the " Natural Law of the Spiritual World,"	MRS. E. B. KENDRICK.

During 1886 the following papers were read:

JANUARY.

The Beginnings of Life, . REV. DR. STIDHAM, of New Britain.
Review of "The New Theology," . REV. J. H. BRECKINRIDGE.

MARCH.

Electric Motors, . . PROF. HENRY S. PRATT.
Horology, . . . REV. J. G. GRISWOLD.

MAY.

Beginnings of Life, with Stereopticon illus-
trations, . . REV. DR. STIDHAM, of New Britain.

OCTOBER.

A Plea for the Birds, . FRANKLIN PLATT.

DECEMBER.

The Niagara Gorge, . . . REV. J. H. CHAPIN.
India, ... MRS. G. H. McGREW.

During the summer an excursion was made to the Portland Quarries.

The lecture course for 1886–7 was very successful and gave great satisfaction. It comprised lectures by

REV. E. C. BOLLES, Ph. D., on "First Letters of the Alphabet of Life."
DR. ALFRED RUSSELL WALLACE, LL. D., F. R. S., on "The Darwinian Theory."
PROF. ALEXANDER WINCHELL, LL. D., on "Life History of the World."
REV. J. H. CHAPIN, Ph. D., on "An Evening in the Land of the Montezumas."
And a concert by the YOUNG APOLLO CLUB.

The Association has a large collection of geological and other specimens which are not unpacked, for want of sufficient room, and it is to be hoped that before the summer vacation we shall have rooms of our own, where our library and collections can be made more available.

CHAS. H. S. DAVIS,

Secretary.

Report of Treasurer.

MERIDEN SCIENTIFIC ASSOCIATION.

1886.   Dr.

To cash paid Mr. Garton,			. $3.00	
"	"	Dr. Chapin,	8.65	
"	"	Robt. Bowman,	5.33	
"	"	Dr. Davis, .	. . 3.00	
"	"	Mrs. Kendrick,	. 10.40	
		Balance,	. . 157.52	
				$187.90

1886.   Cr.

By Balance from 1885,	. $81.90	
" Memberships, . . .	. 2.00	
" Mrs. Kendrick, Collector,	. 104.00	
		$187.90

A. B. MATHER,

Treasurer.

January 10, 1887.

Report of Curator and Librarian.

I have to report the following additions to the library and cabinet of the Association during the years 1885 and 1886:

Geology of the Comstock Lode and the Washoe District. Becker.

United States Geological Survey. Powell. Fifth Annual Report, 1883-1884.

Reports of Observations of the Total Eclipse of the Sun, August 7, 1869. J. H. C. Coffin.

Bureau of Ethnology. Annual Report. J. W. Powell. 1879-80-81-82.

Powdered Anthracite and Gas Fuel. Scranton Board of Trade, 1886.

Smithsonian Report, 1884.

American Ephemeris and Nautical Almanac, 1886-87-88-89.

Astronomical Papers of the American Ephemeris.

Transactions of the Academy of Science of St. Louis. Vol. 4; No. 3.

Bulletin of the Torrey Botanical Club, 1886.

Proceedings and Transactions of the Natural History Society of Glasgow. Vol. 1; Parts 1-2.

Transactions of the Wyoming Historical and Geological Society. Vol. 2; Part 2.

Journal of the Cincinnati Society of Natural History, 1885.

Bulletin of the Minnesota Academy of Natural Sciences, 1882-83.

Transactions of the Vassar Brothers' Institute, Scientific section. Vol. 2.

Monographs of the United States Geological Survey. Vols. 2 to 12, with maps to accompany Vols. 2 and 3.

Bulletin of the United States Geological Survey. Vols. 1, 2, 3.

Sixteen maps accompanying Report of Forest Trees of North America. Prof. C. S. Sargent.

Dinamica—Quimica. C. F. de Landero and Raoul Prieto. Guadalajara, 1886.

Bulletin of Brookville (Ind.) Society of Natural History. No. 2.

Bulletin of the American Museum of Natural History, New York. No. 7.

ADDITIONS TO CABINET.

Fossil Plants, one specimen.

Geodes, three specimens.

Dendrites, four specimens (one very large).

Calcareous Tufa, two specimens.

Graphite, two specimens.

Epidote, two specimens.

Barytes (Coxcomb), three specimens.

Mica (Muscovite), one specimen.

Mica (Biotite), one specimen.

Moonstone, one specimen.

Petrified Wood, one specimen.

Buhrstone, two specimens.

Native Copper, four specimens.

Pyrites, two specimens.

Limonite, three specimens.

Franklinite, two specimens.

Black Jack, two specimens.

Gold-bearing Quartz, two specimens.

Specimens of Iron, Copper, Manganese, Zinc, Pyrolusite, Psilomelane.

Hematite, large and very fine specimens.

The above minerals are from Northern Michigan.

Petrified Wood from Apache County, Arizona.

Feldspar and Rose Quartz from Portland, Conn.

Metamorphosed Sandstone and Conglomerate, from contact line between Trap and Sandstone, East Rock, New Haven, Conn.

Cedar Wood perforated by borers, two specimens.

The following collection of minerals are from the Black Hills, Dakota, in the vicinity of Deadwood, and were presented to the Association by Wm. H. Pratt, Esq.

Copper Ore from Eagle mine, one specimen.

Copper Ore from St. Joseph mine, one specimen.

Silver Ore from Far West mine, three specimens.
Silver Ore from Seabury Calkins mine, two specimens.
Silver Ore from Gem Hill mine, five specimens.
Silver Ore from Elko mine, three specimens.
Silver Ore from La Plata mine, one specimen.
Gold Ore from Homestake mine, seven specimens.
Gold Ore from Deadwood Terror mine, one specimen.
Gold Ore from De Smelt mine, two specimens.
Gold Ore from Caledonia mine, one specimen.
Lead Carbonate from Medora mine, two specimens.
Iron Ore from Medora mine, two specimens.
Bituminous Coal from Haycreek mine, one specimen.
Galena from Hannibal mine, one specimen.
Tin Ore from Nigger Hill mine, one specimen.
Graphite, one specimen.
Gypsum from Black Hills, one specimen.
Marble from Gilman mine, one specimen.
Salt from Iron Springs, fifty miles west of Deadwood.

ROBERT BOWMAN,

Curator and Librarian.

January, 1887.

PROCEEDINGS.

1885.

—

JANUARY 12, 1885.

Regular meeting of the Association, January 12, 1885. Rev. C. H. Everest, Eli C. Birdsey, Mr. and Mrs. H. B. Allen, Jas. F. Allen, Miss Caro G. Allen, Mr. and Mrs. L. L. Sawyer, Miss Sarah N. Benedict and Miss Hettie Lewis were elected members of the Association.

Miss Emma L. Rice and Miss Hattie A. Linsley were proposed for membership.

Voted, that W. H. Quested be appointed Collector.

The following officers were appointed for the ensuing year:

President—REV. J. H. CHAPIN.

Vice-President—REV. J. T. PETTEE.

Secretary—CHAS. H. S. DAVIS, M. D.

Treasurer—ALBERT B. MATHER.

Curator and Librarian—ROBERT BOWMAN.

Voted, That the Rev. J. H. Breckinridge be appointed Necrologist.

Adjourned to meet Monday evening, February 9, 1885.

FEBRUARY 9, 1885.

Regular meeting of the Association, February 9, 1885. Miss Emma L. Rice and Miss Hattie A. Linsley were elected members of the Association.

Messrs. Seidensticker and Pratt and Mrs. Kendrick were appointed committee on programme for the April meeting.

Voted, That when any director of a section is absent, upon the call of that section for the monthly report, any member of the Association may contribute any fact or incident pertaining to that section, and when any director has completed his monthly report, any member present may supplement it by any fact or incident within his knowledge.

Adjourned to meet Monday evening, March 9, 1885.

MARCH 9, 1885.

Regular meeting of the Association, March 9, 1885. E. C. Wheatley and Miss M. H. Nash were proposed for membership.

Voted, That the number of copies of the TRANSACTIONS to be printed be left with the committee.

Adjourned to meet Monday evening, April 13, 1885.

APRIL 13, 1885.

On account of the Hospital Fund concert, the April meeting of the Association was postponed until May 11, 1885.

MAY 11, 1885.

Regular meeting of the Association, May 11, 1885.

E. C. Wheatley and Miss M. H. Nash were elected members of the Association.

Messrs. Hirons and Beebe were appointed committee on programme for the September meeting.

Voted, That the Association have an excursion some Saturday before the end of the school year, and Messrs. Bowman, Seidensticker and Pratt were appointed a committee to select the place and make the necessary arrangements.

Adjourned, to meet Monday evening, June 8, 1885.

JUNE 8, 1885.

Regular meeting of the Association, June 8, 1885. Newell Wightman and D. S. Carpenter, of West Rutland, Vt., were proposed for membership.

Voted, That the officers of the Association be appointed a committee to consider what can be done to extend the usefulness of the Association. Adjourned to meet Monday evening, September 14, 1885.

SEPTEMBER 14, 1885.

Regular meeting of the Association, September 14, 1885. D. S. Carpenter, of West Rutland, Vt., and Newell Wightman, of South Meriden, were elected members of the Association.

Messrs. Davis, Mather and Pratt were appointed a committee to purchase books for the Association.

Messrs. Davis and Barker and Miss Daniels were appointed a committee on programme for the November meeting of the Association.

Adjourned to meet Monday evening, October 12, 1885.

OCTOBER 12, 1885.

Regular meeting of the Association, October 12, 1885. Thomas L. Reilly, William M. Stoddard, George L. Cooper, Mrs. Ella Hood Cooper, Rev. C. A. Knickerbocker, Mrs. George Austin Fay, Mrs. Mary E. Pratt, Rev. Isaac R. Wheelock, Miss Alice Derby and Miss Addie Upson were proposed for membership.

Voted, That Mrs. Kendrick purchase sufficient paper for mounting the Herbarium.

Adjourned to meet Monday evening, November 10, 1885.

NOVEMBER 10, 1885.

Regular meeting of the Association, November 10, 1885. T. S. Reilly, W. L. Stoddard, Mr. and Mrs. G. L. Cooper, Rev. C. A. Knickerbocker, Rev. I. R. Wheelock, Mrs. G. A. Fay, Mrs. Mary E. Pratt, Miss Alice Derby and Miss Addie Upson were elected members of the Association.

Rev. Mr. Markwick was proposed for membership.

Messrs. Garton and Stone and Mrs. Kendrick were appointed a committee on programme for the January meeting.

Adjourned to meet Monday evening, December 14, 1885.

DECEMBER 14, 1885.

Regular meeting of the Association, December 14, 1885. Rev. Mr. Markwick was elected a member of the Association. Adjourned to meet Monday evening, January 11. 1886.

1886.

JANUARY 11, 1886.

Regular meeting of the Association, January 11, 1886. Messrs. Mather, Robinson and Upson were appointed a committee on programme for the March meeting.

Miss Wegia Hope Hall was proposed for membership.

Voted, That the bills presented by Rev. Dr. Chapin to the amount of $13.30 be paid.

Rev. Dr. Stidham was proposed for honorary membership.

Mrs. E. B. Kendrick was appointed Collector for the ensuing year.

The following officers were appointed for the ensuing year:

President—REV. J. H. CHAPIN.

Vice-President—REV. J. T. PETTEE.

Secretary—DR. C. H. S. DAVIS.

Treasurer—A. B. MATHER.

Curator and Librarian—ROBERT BOWMAN.

Voted, That the directorships remain as the preceding year.

· Adjourned to meet Monday evening, February 8, 1886.

FEBRUARY 8, 1886.

Regular meeting of the Association, February 8, 1886.

Miss Wegia H. Hall was elected an active member of the Association.

Rev. Dr. Stidham was elected an honorary member of the Association. Benjamin C. Kennard and Mrs. Josephine Prevost were proposed for membership.

Voted, That a committee of three be appointed by the President of the Association to wait on Rev. J. T. Pettee and request him to withdraw his resignation.

Messrs. Davis, Mather and Bowman were appointed as the committee.

Adjourned to meet Monday evening, March 8, 1886.

MARCH 8, 1886.

Regular meeting of the Association, March 8, 1886. Benjamin C. Kennard and Mrs. Josephine Prevost were elected to membership.

Dr. Anna Jackson Ferris, H. O. Winslow, Mrs. Caroline Winslow, Franklin Platt, Mrs. Sarah R. Platt, Mrs. Clarence E. Ellsbree and Rev. J. G. Griswold were proposed for membership.

Messrs. Garton and Stone and Mrs. Kendrick were appointed a committee on programme for the May meeting.

Voted, that a committee be appointed by the President to consider the advisability of establishing a department of social science, to report at the April meeting of the Association.

Messrs. Hall, Davis and Pratt were appointed as the committee.

Adjourned to meet Monday evening, April 12, 1886.

APRIL 12, 1886.

Regular meeting of the Association April 12, 1886. Dr. Anna Jackson Ferris, H. O. Winslow, Mrs. Caroline Winslow, Franklin Platt, Mrs. Sarah R. Platt, Mrs. Clarence E. Ellsbree and J. G. Griswold were elected to membership.

The committees on programme and on social science were continued to the May meeting of the Association.

Adjourned to meet Monday evening, May 10, 1886.

MAY 10, 1886.

Regular meeting of the Association, May 10, 1886.

Messrs. Pettee, Davis and Quested were appointed a committee to arrange for an excursion of the Association.

Messrs. Mather, Cooper, Bowman and Pratt and Mrs. Kendrick were appointed a committee to consider the expediency of having a course of lectures the coming winter.

Messrs. Pratt and Stone and Miss Ida Hall were appointed a committee on programme for the September meeting.

Adjourned to meet Monday evening, June 14, 1886.

JUNE 14, 1886.

Regular meeting of the Association, June 14, 1886. David S. Root and Rev. S. Halsted Watkins were proposed for membership.

Report of committee on excursion accepted and the committee directed to arrange for the excursion to Portland.

Voted, that the lecture committee proceed to arrange for a course of lectures and to report at the September meeting.

Voted, That the September meeting be devoted to the reports of the directors of sections.

Voted, That Mr. Franklin Platt be appointed director of the section of Ornithology.

Adjourned to meet Monday evening, September 13, 1886.

SEPTEMBER 13, 1886.

Regular meeting of the Association, September 13, 1886.

Report of committee on programme accepted and committee discharged. Rev. George H. McGrew, S. J. Robie, Mr. and Mrs. Frank Rhind, Mrs. Knickerbocker and Miss F. Ione Garde were proposed for membership.

Adjourned to meet Monday evening, October 11, 1886.

OCTOBER 11, 1886.

Regular meeting of the Association, Monday evening, October 11, 1886. Rev. George H. McGrew, S. J. Robie, Mr. and Mrs. Frank Rhind, Mrs. Knickerbocker and Miss F. Ione Garde were elected to membership.

Miss Annie Bishop, Augustus Hirschfeld, David Mossman and Wm. W. Lee were proposed for membership.

Messrs. Platt and Everitt and Miss Moess were appointed committee on programme for the November meeting.

Voted, that the lecture committee confer with Mr. Marvin in regard to selling reserved seats for the lecture course.

Adjourned to meet Monday evening, November 8, 1886.

NOVEMBER 8, 1886.

Regular meeting of the Association, November 8, 1886. Miss Annie Bishop, August Hirschfeld, David Mossman and Wm. W. Lee were elected to membership.

Mr. and Mrs. Nathan S. Baldwin, Frank E. Sands, O. G. Harrison, Charles G. Kendrick and Mrs. Carrie E. Davis were proposed for membership.

Committee on programme reported and the committee continued.

Adjourned to meet Monday evening, December 6, 1886.

DECEMBER 6, 1886.

Regular meeting of the Association, December 6, 1886.

Messrs. C. E. Ellsbree and Stoddard and Mrs. Everitt were appointed committee on programme for the February meeting.

Messrs. Davis and Platt and Mrs. Kendrick were appointed committee on publication.

Adjourned to meet Monday evening, January 10, 1887.

CHAS. H. S. DAVIS,

Secretary.

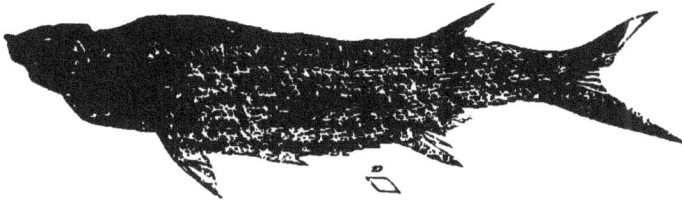

The Catopterus gracilis.

BY CHAS. H. S. DAVIS, M. D.

There is an enormous series of subaqueous sediment, originally composed of mud, sand or pebbles, the successive bottoms of a former sea, and in which no trace of organic life has yet been detected.

These non-fossiliferous sedementary beds form, in all countries where they have been examined, the base-rock on which the Cambrian or oldest Silurian strata rest.

Whether they be significative of ocean abysses never reached by the remains of coeval living beings, or whether they truly indicate the period antecedent to the beginning of life on this planet, are questions of the deepest significance, and demanding much farther observation before they can be authoritatively answered.

Palæontologists fail to detect a single bone of any aquatic animal of the vertebrate class in rocks older than the uppermost division of the Silurian system, but fishes are found in all the great rock foundations from the graywacke upward, and, therefore, the history of fossil fishes becomes of great importance.

When we consider how rich a molluscuous fauna, to say nothing of the crustaceans, sea urchins, corals, etc., which have been met with in almost all parts of the world, it seems impossible to account for our not having yet found any accompanying bones of fish, except by supposing they were not yet in being, or that they only occupied a limited area.

Next to the Silurian comes the Old Red-sandstone Devonian formation, which is very rich in fishes, the majority of which be-

[The Association is indebted to Messrs. Ivison, Blakeman & Taylor, the publishers of *Dana's Geology*, for the use of the cut of the *Catopterus gracilis*.]

longed to the order of *Ganoids* and a few to the *Placoids* of Agassiz; and it is remarkable that the vast majority of the fossil fish of the succeeding formation, from the Carboniferous to the Oolitic, consists, in like manner, of *Ganoids*, a family which, though so rich in genera in the olden times, is of quite exceptional occurrence in the present creation.

These fish belong to races, of which there have been left as their living representatives on the globe, only two genera and about seven species, viz., the *Lepidosteus*, or bony fishes of this country and the *Polypterus* of the Nile and Senegal, although seventeen genera have been found fossil.

The present might appear to be the culminating period in the development of fishes, in respect to the number of ordinal forms or modifications of the class.   It represents, however, rather the result of mutation, depending upon the progressive assumption of a more special type.

The cabinet of the Meriden Scientific Association contains some very fine specimens of the fossil fish known as the *Catopterus gracilis*, which were procured by members of the Association at a place called Little Falls, about eight miles east of Meriden and two miles north of Durham Center.

Fossil fish have been found in several places in Connecticut and at Sunderland, Deerfield and West Springfield, Mass.; in fact, with the exception of the teeth and vertebræ of sharks, found in the cretaceous formation of the Alantic coast, the fossil remains of fishes hitherto discovered in the United States have, for the most part, been confined to the New Red-sandstone of the Connecticut river valley.

These fishes are in most cases found in bituminous shale, which, in character, sometimes approaches a micaceous sandstone.   The shale generally forms the bank of the river several feet high and the ichthyolites are most abundant in the lower part of the bed which corresponds nearly with low water mark.   A thin layer of carbonaceous matter usually marks out the spot where the fish lay, except the head, whose outlines are rendered visible only by irregular ridges and furrows.   In some cases, however, satin spar forms a thin layer over the carbonaceous matter, and being of a light grey color, it gives to the specimen an aspect extremely like that of a fish just taken from the water.

The *Catopterus gracilis* was first described by Redfield.*

---

*Annals of the Lyceum of Natural History, New York, 1837; p. 35.

In the arrangement of Agassiz* the *Catopterus gracilis* would be comprehended in the order of *Ganoids* and family *Lepidoides.* Redfield was disposed at first to assign it to the Homocercal division of Agassiz's family *Lepidoides,* but was afterwards disposed to qualify this somewhat, and judged it to occupy a sort of intermediate position between the two divisions, neither being exactly equilobed, like the *Homocerci,* nor yet having the decided heterocercal character which belongs to those genera which Agassiz has placed in that division.

The name *Catopterus* was given to this genus by Redfield, from the situation of the dorsal fin.

At Little Falls, the bituminous shale in which the fishes are found, occurs interstratified with the sandstone, and is exposed to view at the bottom of a ravine twenty or thirty feet in depth, which has been excavated by the action of a small stream. The stratum is nearly horizontal. Some layers of the shale abound, not only in remains of fishes, but also in those of vegetables, apparently endogenous, while others are nearly destitute of both.

The substance of the fish, as well as that of the vegetable, is converted into carbonaceous matter, and it is observed that while the form of the scales and rays is perfectly and beautifully preserved, there are no traces of bones remaining.

According to Agassiz, this is almost universally the case with the individuals of the family *Lepidoides* to which these belong.

The following is a description of the *Catopterus gracilis*:

Body, fusiform, covered with rhomboidal scales, which extend obliquely across it, and parallel with its length. *Scales,* middling size. *Head,* rather small, presenting a finely granulated surface, resembling shagreen. *Back,* but slightly arched. *Pectoral fins,* middling. *Ventral,* small, inserted midway between pectoral and anal. *Anal,* large. *Dorsal,* middling, extended opposite the posterior part of the anal. *Tail,* forked, equilobed. *Scales,* extending a little upon the base of the upper lobe. All of the fins have a se-

---

* Agassiz divides fishes into four orders, deriving their characters from the scales: 1. The *Placoidians,* or those whose skin is covered irregularly with plates of enamel. 2. The *Ganoidians,* or those having angular scales of horny or bony plates, covered with a thick plate of enamel. 3. The *Ctenoidians,* or those having jagged or pectinated scales. 4. The *Cycloidians,* or those having scales smooth and simple at their margin. Three-fourths of the existing species of fish belong to the two last orders, whose existence has not been ascertained below the chalk. The remaining fourth belongs to the two first orders, and existed alone in all the periods during which the fossiliferous rocks below the chalk were deposited. *Agassiz: American Journal of Science, Vol. 30; p. 39.*

ries of raylets inserted obliquely upon the first or anterior ray, producing a serrated or denticulate appearance. The succeeding rays
have an articulate appearance and are finely subdivided toward
their extremities. The following list of the number of rays in each
fin may serve to give an idea of their relative size:

Pectoral, large and strong. . . . . 10 to 12.
Ventral . . . . about 8.
Dorsal . . . . 10 to 12.
Anal . . . . 20 to 30.
Caudal . . . . 30 to 40.

The Hanging Hills.

By J. H. Chapin, Ph. D.

A fresh interest attaches to the Hanging Hills—the trap ridges lying just out of the city limits in Meriden, toward the west—in the discussion now attracting attention in regard to their origin, or rather the mode of their occurrence.

In the Triassic, or Red-sandstone region of the Connecticut valley—extending from the northern limits of Massachusetts to Long Island Sound—are numerous ridges of a dark colored eruptive rock, somewhat indefinitely described as trap, that have sometime in the past history of this region been pushed up through rifts in the red sandstone for which the section is so well-known.

No one passing through Meriden, even on the railroad, can fail to observe them, so prominent and striking a feature are they of the landscape.

And indeed in going from New Haven to Greenfield, Mass., from one to half a dozen of these singular prominences are always in sight.

Though seemingly disposed at irregular intervals, the map shows that they form a continuous series, not unlike a mountain chain, and suggest at once the presence of some of the forces that from time to time have thrown the earth's surface into rugged folds. But a casual examination is sufficient to show that the action here was different from that in ordinary mountain making.

The black rock mass is no part of the original strata laid down in seas or rivers, but an intruder, that found its way hither after the Triassic formation had been deposited, and for which the strata had been rudely disturbed and torn asunder.

The term Triassic is here used, as the reader will understand, to include the Jurassic also, or so much of it as may be represented along the the eastern border of the continent. Perhaps Tria-Jurassic, or as Le Conte puts it, Jura-Triassic, would be more appropriate.

The range of trap ridges, extending from Mounts Tom and Holyoke in Massachusetts on the north, has a southerly trend, nearly parallel to the Appalachian chain, till it reaches the Hanging Hills, buttressed and sentineled by West Peak, in Meriden, where it makes an abrupt turn to the east along West Mountain, South Mountain, Cathole aud Mount Lamentation, when it bends southward again through Middlefield, Durham, and so on, and encloses the city of Meriden like the ramparts of a vast amphitheatre. An almost invariable feature of these ridges is that one side is abrupt, and the other a moderate slope, the latter indicating the direction of the flow from the elongated vent or crater. When the direction of the ridge is from north to south the abrupt side is on the west, and when the direction is from east to west the abrupt side is on the south; thus all the ridges around Meriden present a sort of mural front to the city.

The Hanging Hills are seen to best advantage from the road that follows the ridge south of Crow Hollow (why not call it Ravendale), and crosses what is commonly known as Johnson's Hill. From this outlook the bold, serrated ridge confronts the observer on the north ; South Mountain and West Mountain, separated by a narrow valley now appropriated by the city as a reservoir for its water supply. The most westerly of these uplifts, West Mountain, is riven at one point almost to its base by a gorge running north and dividing the ridge into two sections, the eastern and the larger one simulating a vast ruin with partially shattered walls and crumbling towers, while the western section forms a bold prominence nearly a thousand feet in height, being the highest of the series in Connecticut. South Mountain, east of the reservoir, is cleft into three sections by rifts in the rocky mass and gradually rises from its eastern border to its western summit.

Still eastward is Cathole Mountain, so called from a narrow defile that separates it from that just described. This defile, as it was in a state of nature, suggested the famous pass of Thermopylæ, where, "between the mountain on the one hand and the morass upon the other," but one man could pass at a time. But the demands of trade and travel have led to its enlargement and the highway from Meriden to Kensington now passes through it, making one of the most strikingly picturesque drives in Connecticut.

These ridges have mostly been described as trap dikes, though for reasons that will by and by appear, the term is by some geologists considered inappropriate and misleading. That they are made up entirely of eruptive rocks, however, that have sometime been pushed up through fissures in the sandstone in a heated and plastic condition, there is no chance for doubt. That the mass was at first plastic, or of a partially molten character, is evinced by the fact that on coming to the surface it often flowed away from the vent in one or more directions, thus forming a kind of sheet covering a considerable extent of the adjacent surface.

And that it was heated to a high degree is evidenced by the metamorphism of the sandstone or other contiguous rock, both above and below the sheet, and on either side of the rift through which the mass protruded. The metamorphism or change consists sometimes merely in the induration of the adjacent rocks and sometimes in their more or less complete crystalization ; the sandstone in some instances assuming a texture and hardness that make it scarcely distinguishable from the trap or eruptive matter. Moreover, while the trap seemed to heat the sandstone, the sandstone in turn seemed to cool the trap more rapidly, and so to change it both in color and texture. It is sometimes difficult to determine the precise line of juncture on account of the change that both kinds of rock have undergone, though at the distance of a few feet, possibly a few inches, in either direction, the distinction is very plain. There are examples of contact of the trap with the sandstone exposed by the recent cutting of the drives up and down East Rock, near New Haven, which belongs to this same series ; also a fine example where a vein of trap cuts across the sandstone at Whitneyville. Other examples are described in South Hadley, Massachusetts and at other points, all showing essentially the same characteristics, though there are no very satisfactory exposures as yet in Meriden, the broad and deep slopes of *debris* along the base of the cliffs effectually concealing the line of contact.

But to return to the interesting discussion of which we spoke in the outset. The commonly accepted theory, as to the mode of occurrence of these trap ridges, is that after the sandstone was laid down—or say about the close of the Triassic period—by some profound disturbance of the earth's crust, the whole series of rocks was fractured at various points, and through the fissures thus produced, the trap was pushed to the surface, sometimes only filling the rift like a vein and sometimes overflowing to greater or less extent, and in that condition cooling and hardening and producing the masses as they exist to-day, except that the gradual erosion, or wearing away of the softer sandstone, has served to make them more and more prominent as the ages went by.*

---

* We make no attempt in this paper to distinguish between the older and more

While this theory supposes the several fractures in the series to have been due to the same or similar causes, yet each may have been independent of the others, or only dependent on each other, as one mountain may be dependent upon adjacent mountains. They may have been formed contemporaneously or in succession, following, however, a somewhat definite order and conforming to some definite outline. That is to say, East Rock and Mt. Holyoke may have been formed at the same time, or at periods somewhat remote from each other. Or, to confine the illustration to a narrower range, West Peak and Cathole Mountain may have been pushed up through the sandstone at the same time, or one somewhat in advance of the other. But the rents or fissures were practically independent of each other, and so far as appears now, the one might have been without the other. In other words, the hundred or more ridges that Percival represents on his map of the region presupposes a hundred or more rents in the earth's crust, many of them very intimately related no doubt, and still a hundred separate vents. The other theory as to the mode of occurrence of the trap ridges, and one which was presented by Prof. W. M. Davis, of Harvard University, in a somewhat elaborate paper at the meeting of the American Association for the Advancement of Science at Buffalo, August, 1886, may be briefly stated in the author's own words, as follows—"that nearly all these ridges are the outcropping edges of contemporaneous lava overflows, and that the outcrop of a single sheet is repeated several times by faults nearly paralled to the strike of the beds."

A detailed statement of the evidence bearing on the subject is promised in the forthcoming Seventh Annual Report of the U. S. Geological Survey, for which we shall look with interest. But as we understand the theory, it may be briefly stated thus. The overflow of the trap, whether from few or many vents, formed a continuous and essentially horizontal sheet over a large portion of the area now included in the trap region, *before the occurrence* of the ruptures and dislocations that produced the striking features so apparent now : while the separate ridges, commonly supposed to be due to separate eruptions, are the result of profound fractures and *faulting* of the whole series of rocks, including not only the sandstone and the trap, but the crystalline rocks that lie at still

recent traps, or between that still buried in the sandstone and that exposed upon the surface. This distinction is made by Prof. W. N. Rice in a recent article— Am. Jour. of Science, December, 1886—and illustrated by reference to some exposures in the vicinity of Tariffville, Conn.

greater depth. Thus the overflow of the trap and the erection into ridges were separate events, and may have been wide apart in point of time. There are confessedly exceptions to the rule, in the the case of sheets or layers of trap that were sometimes intruded between successive layers of the sandstone, in which case it came to the surface at an angle corresponding to the dip of the strata and may or may not have overflowed the surface. The one theory makes the the forces that produced the original fractures in the crust and gave vent to the molten matter, the same that built up the ridges; the other makes them independent of each other, even though they may have been of the same general character and of similar origin. It makes West Peak and Cathole and Mount Lamentation portions of a common trap sheet, merely separated by great faults, and even recognizes subordinate faults along the clefts that partially subdivide South Mountain and others of like configuration.

It is too soon to attempt a statement of the comparative evidence bearing upon these two theories, and there are problems involved that may well delay the settlement of the question for many a day. The statement of two or three points, however, may help to place the matter fairly before the reader.

1. It has long appeared that that there is a general dip of the sandstone of Connecticut toward the east. It is also in evidence that the trap, in part at least, lies conformably upon the sandstone, and has therefore an easterly dip. One theory would make this dipping of the sandstone the result of an uplift of the whole area along its western border, thus tilting it to the east. And as the dikes or veins of trap have penetrated the strata thus tilted, the presumption is that the eruptions occurred subsequent to the tilting, or possibly contemporaneous with it, the two operations being due to the same general cause.

The other theory assumes that the whole formation, including the sandstone and the eruptive trap, was completed, or at least well advanced, before the disturbance occurred which produced the ridges; and that the latter are due not to fresh ejections of material, but to faulting of the rocks already in place, the downthrow being on the west side of the fracture, thus producing a dip or inclination to the eastward, but each section of faulted area being independent in the position it assumes, and the dip therefore not necessarily uniform. If the sandstone formed but a thin stratum, this question would be easily decided, for the upthrow of any great fault would expose the entire thickness. But the thickness of the formation is too great for reliable evidence on this point.

2. Another point worthy of consideration is that if the trap existed as a continuous sheet, it is reasonable to suppose that the rock would be uniform in character, or that the changes in chemical composition, if not in structure, would be gradual rather than abrupt. This does not appear to be the case. The rock differs not only in composition but in structure on opposite sides sometimes of a narrow ravine, to whatever cause that ravine may be due. That a variety should exist is nothing strange, but some cause for the variety might reasonably be expected. Sometimes we find the trap-rock to be homogeneous, and sometimes amygdaloidal; sometimes very compact, sometimes loosely aggregated, and again assuming the porous or vesicular character antecedent to amygdaloidal deposits. And if it be true that amygdaloidal deposits usually occur near a cooling surface, this would not account for the presence of amygdaloid in one ridge and its absence in an adjacent ridge, *if* the existence of the two was due merely to a fracture and fault; for the cooling must have taken place before the faulting occurred.

Altogether, the problem is a perplexing one and whatever may come of the discussion, the effect will be to awaken new interest and make us more familiar with these unique formations in our own vicinity.

An Interesting Find.

On a recent excursion to the quarries in Durham, which have yielded from time to time so many interesting specimens of the *fauna* of the Triassic rocks of Connecticut, H. H. Kendrick, a member of the Meriden Scientific Association, brought away several fossil specimens of the *flora* of that interesting period. The first bears some resemblance to the fruit of the early conifers, but is more likely that of a cycad—probably *cycadites*—having a rounded base and being pointed at the other extremity, with a structure more nearly resembling a bud than a well defined nut. Another piece of shale contains two impressions of a similar form.

Another specimen is that of a plant akin to the calamite, though with joints imperfectly defined ; probably one of the dwarfed varieties that followed the period of the great coal plants.

Another bit of shale contained small impressions of a similar plant, though too fragmentary to be determined with any certainty ; while a small slab showed the wave or ripple marks indicative of a shallow sea.

J. H. C.

A List of the Birds of Meriden, Conn.

BY FRANKLIN PLATT.

With the mountains and wooded hills which constitute the northern and eastern boundaries of Meriden, with the waters of the Quinnipiac running through her borders, with the numerous natural and artificial ponds—some of which are of no mean size—which repose here and there in her valleys, and her alternating stretches of meadow and woodland and an occasional piece of swamp-ground, it would seem as if nature had provided her with all the variety of physical features necessary to accommodate every species of bird likely to be found in any inland town of the State. And if the following list shall be found to be incomplete—if the names of some birds liable to be found here have been omitted—the chances, in my opinion are, that the omissions have occurred, not because of the continuous absence of such species, but because it has not been my good fortune either to discover their presence or to obtain satisfactory evidence of their occurrence.

As my studies of bird life have been so much confined to those birds more generally found in our fields and woods, I have had to rely to a considerable extent on such information as I could get from other persons as to the occurrence of many of the water birds on my list and some of the birds of prey. But in no such case even have I relied on the mere conjecture of any person. And where I have had doubts in the matter I have either omitted the name of the doubtful species entirely, or have included it in the list with such observations as tend to show the probabilities of its occurrence.

And right here I desire to acknowledge the indebtedness I am under to Mrs. Nathan S. Baldwin—by referring to her collection of stuffed birds I have had my doubts removed in more than one instance; to the brothers George and Charles I. Foster and to Franklin T. Ives, Esq., men of considerable experience as sportsmen in this vicinity, all of whom have given me valuable information on important points.   I am also under obligations to Mr. J. A. Allen, the Curator for the Department of Ornithology in the American Museum of Natural History, at New York, for some useful hints in regard to the classifying, etc., of the birds herein named.

The nomenclature used is taken from The Code of Nomenclature adopted by the American Ornithologists' Union.*   In some cases I have added in parenthesis the locally more common names.

Where I have mentioned a species as *resident* I have not meant that the same individual birds of that species stay with us the year through the same as our summer birds which are marked as summer residents do through the summer months, but that birds of that species are to be found here throughout the year.   Again, in noting a species as *common*, I wish the word to be understood in a comparative sense, particularly when applied to water-birds and birds of prey, meaning that such species are common as compared with others of the same family which are much less common.   As a rule, too, most of the species mentioned as common are so only in favorable localities.   In some cases I have seen fit to give, in the interest of readers in this vicinity, the localities where certain species may be found.

— .. —

Order PYGOPODES.   Diving Birds.

Suborder PODICIPEDES.   Grebes.

Family PODICIPIDÆ.   Grebes.

Genus PODILYMBUS Lesson.

1.   **Podilymbus podiceps** (Linn.).
    Pied-billed Grebe.   (Dipper.   Hell-Diver.)
Rather common during the spring and fall migrations.

Suborder CEPPHI.   Loons and Auks.

Family URINATORIDÆ.   Loons.

Genus URINATOR Cuvier.

* The Code of Nomenclature and Check List of North American birds adopted by the American Ornithologists' Union, being the Report of the Committee of the Union on Classification and Nomenclature.   New York, 1886.

2. **Urinator imber** (GUNN.).

Loon.

Rare visitant. (Has been seen late in the fall of the year at the Foster Brothers' Pond and at Black Pond.)

ORDER ANSERES. LAMELLIROSTRAL SWIMMERS.

FAMILY ANATIDÆ. DUCKS, GEESE AND SWANS.

SUBFAMILY MERGINÆ. MERGANSERS.

GENUS LOPHODYTES REICHENBACH.

3. **Lophodytes cucullatus** (LINN.).

Hooded Merganser.

SUBFAMILY ANATINÆ. RIVER DUCKS.

GENUS ANAS LINNÆUS.

4. **Anas boschas** LINN.

Mallard.

Uncommon visitant. (Specimens have been taken at Foster Brothers' Pond.)

5. **Anas obscura** GMEL.

Black Duck.

Rather common summer resident, but most numerous in the fall and spring.

SUBGENUS MARECA STEPHENS.

6. **Anas Americana** GMEL.

Baldpate.

Rare spring and autumn visitant.

SUBGENUS NETTION KAUP.

7. **Anas carolinensis** GMELIN.

Green-winged Teal.

Not uncommon during the migrations, particularly in the fall.

SUBGENUS **QUERQUEDULA** STEPHENS.

8. **Anas discors** LINN.

Blue-winged Teal.

Rather common migrant, especially in the fall.

GENUS **AIX** BOIE.

9. **Aix sponsa** (LINN.).

Wood Duck.

Rather common summer resident.

GENUS **CHARITONETTA** STEJNEGER.

10. **Charitonetta albeola** (LINN.).

Bufflehead.

Occasional visitant.

SUBFAMILY **ANSERINÆ.** GEESE.

GENUS **BRANTA** SCOPOLI.

11. **Branta canadensis** (LINN.).

Canada Goose.

Not uncommon visitant during the migrations. Black Pond, North Farms Reservoir.

12. **Branta bernicla** (LINN.).

Brant.

Very rare visitant. The only instance of its occurrence which has come to my notice was that of four years ago (1882), when Mr. Charles I. Foster captured two out of a small flock that had stopped at the Foster Brothers' Pond.

ORDER HERODIONES. HERONS, STORKS, IBISES, ETC.

SUBORDER HERODII. HERONS, EGRETS, BITTERNS, ETC.

FAMILY **ARDEIDÆ.** HERONS, BITTERNS, ETC.

SUBFAMILY **BOTAURINÆ.** BITTERNS.

GENUS **BOTAURUS** HERMANN.

SUBGENUS **BOTAURUS.**

13. Botaurus lentiginosus (Montag.).

American Bittern.

Probably an occasional summer resident. Mr. F. T. Ives says it was quite common some years ago around the North Farms Reservoir.

SUBFAMILY **ARDEINÆ.** Herons and Egrets.

GENUS **ARDEA** Linnæus.

SUBGENUS **ARDEA.**

14. Ardea herodias Linn.

Great Blue Heron.

A not uncommon visiter in spring and fall.

SUBGENUS **HERODIAS** Boie.

15. Ardea egretta Gmel.

American Egret. (Great White Heron.)

Very rare visitant. Has been seen at Black Pond and at Foster's Pond. Mrs. Baldwin has a fine specimen of this elegant bird, which was taken a few years ago at the North Farms Reservoir.

SUBGENUS **BUTORIDES** Blyth.

16. Ardea virescens Linn.

Green Heron.

A not uncommon summer resident.

GENUS **NYCTICORAX** Stephens.

SUBGENUS **NYCTICORAX.**

17. Nycticorax nycticorax nævius (Bodd.).

Black-crowned Night Heron. (Quawk.)

Rather common summer resident.

ORDER **PALUDICOLÆ.** Cranes, Rails, Etc.

SUBORDER RALLI. Rails, Gallinules, Coots, Etc.

FAMILY **RALLIDÆ.** Rails, Gallinules and Coots,

SUBFAMILY **RALLINÆ.** Rails.

GENUS **PORZANA** Vieillot,

SUBGENUS **PORZANA.**

18. **Porzana carolina** (LINN.).

Sora.

Specimens have been taken in this vicinity, but I cannot learn definitely whether they ever breed here or not.

SUBFAMILY **FULICINÆ.** COOTS.

GENUS **FULICA** LINNÆUS.

19. **Fulica americana** GMEL.

American Coot.

Occasional visitant, mostly in the autumn.

ORDER LIMICOLÆ. SHORE BIRDS.

FAMILY **SCOLOPACIDÆ.** SNIPES, SANDPIPERS, ETC.

GENUS **PHILOHELA** GRAY.

20. **Philohela minor** (GMEL.).

American Woodcock.

Rather common summer resident. More numerous in spring and fall during the migrations, particularly in the fall.

GENUS **GALLINAGO** LEACH.

21. **Gallinago delicata** (ORD.).

Wilson's Snipe. (Known as "English Snipe," "Jack Snipe," etc.)

More or less common during the migrations.

GENUS **TOTANUS** BECHSTEIN.

SUBGENUS **GLOTTIS** KOCH.

22. **Totanus melanoleucus** (GMEL.).

Greater Yellow-legs.

Not uncommon during the migrations.

23. **Totanus flavipes** (GMEL.).

Yellow-legs. (Smaller Yellow-legs.)

Only during the migrations. More common in the fall. Both this and the preceding species were quite common around the North Farms Reservoir some years ago.

Genus **BARTRAMIA** Lesson.

24. **Bartramia longicauda** (Bechst.).

    Bartramian Sandpiper. (Upland Plover. Grass Plover.)

Not uncommon summer resident. Common during the migrations.

Genus **ACTITIS** Illiger.

25. **Actitis macularia** (Linn.).

    Spotted Sandpiper. (Tip-up. Teeter-tail. Peet-weet.)

Common summer resident.

Order GALLINÆ. Gallinaceous Birds.

Suborder PHASIANI. Pheasants, Grouse, Partridges, Quails, Etc.

Family TETRAONIDÆ. Grouse, Partridges, Etc.

Subfamily PERDICINÆ. Partridges.

Genus **COLINUS** Lesson.

26. **Colinus virginianus** (Linn.).

    Bob-white. (Quail.)

Common resident.

Subfamily TETRAONINÆ. Grouse.

Genus **BONASA** Stephens.

27. **Bonasa umbellus** (Linn.).

    Ruffed Grouse. (Almost always spoken of in this vicinity as *Partridge*.)

Common resident.

Order COLUMBÆ. Pigeons.

Family COLUMBIDÆ. Pigeons.

Genus **ECTOPISTES** Swainson.

28. **Ectopistes migratorius** (Linn.).

    Passenger Pigeon. (Wild Pigeon.)

Only during the fall migrations and then not common. From thirty to fifty years ago they were very plenty every fall.

Genus ZENAIDURA Bonaparte.

29. Zenaidura macroura (Linn.)

Mourning Dove.

Summer resident. Not uncommon.

Order RAPTORES. Birds of Prey.

Suborder FALCONES. Vultures, Falcons, Hawks, Buzzards, Eagles, Kites, Harriers, Etc.

Family FALCONIDÆ. Vultures, Falcons, Hawks, Eagles, Etc.

Subfamily ACCIPITRINÆ. Kites, Buzzards, Hawks, Goshawks, Eagles, Etc.

Genus CIRCUS Lacépède.

30. Circus hudsonius (Linn.).

Marsh Hawk.

Rather common in spring, summer and fall. Undoubtedly breeds here more or less every spring.

Genus ACCIPITER Brisson.

Subgenus ACCIPITER.

31. Accipiter velox (Wils.).

Sharp-shinned Hawk. (This species is generally known here as the Pigeon Hawk, as well as the true Pigeon Hawk, *Falco columbarius.*)

More or less common summer resident.

32. Accipiter cooperi (Bonap.).

Cooper's Hawk. (Chicken Hawk.)

Common summer resident.

Genus BUTEO Cuvier.

33. Buteo borealis (Gmel.).

Red-tailed Hawk. (Hen Hawk.)

Known to occasionally breed here. More numerous in fall and spring and sometimes seen in winter.

34. Buteo lineatus (Gmel.).

Red-shouldered Hawk.

More common than the preceding species throughout the year.

GENUS **HALIÆETUS** SAVIGNY.

35.   Haliæetus leucocephalus (LINN.).
Bald Eagle.

I am informed that a pair of these birds bred a few years ago in the vicinity of West Peak. Thirty years or more ago a nest was found and the eggs secured on the mountains southeast of the town and the male bird was shot. A pair has also been seen several times during the past spring and summer in the eastern part of the town, which probably nested and raised their young on the mountains there.

SUBFAMILY **FALCONINÆ**. FALCONS.

GENUS **FALCO** LINNÆUS.

SUBGENUS **ÆSALON** KAUP.

36.   Falco columbarius LINN.
Pigeon Hawk.

Occasional in spring, fall and winter.

SUBGENUS **TINNUNCULUS** VIEILLOT.

37.   Falco sparverius LINN.
American Sparrow Hawk.

Rather common every year.

SUBFAMILY **PANDIONINÆ**. OSPREYS.

GENUS **PANDION** SAVIGNY.

38.   Pandion haliaëtus carolinesis (GMEL.).
American Osprey.   (Fish Hawk.)

Uncommon visitor.

SUBORDER **STRIGES**. OWLS.

FAMILY **BUBONIDÆ**. HORNED OWLS, ETC.

GENUS **ASIO** BRISSON.

39.   Asio wilsonianus (LESS.).
American Long-eared Owl.

Resident.   Rather rare.

40.   Asio accipitrinus (PALL.).
Short-eared Owl.

Resident.   Perhaps more common than the preceding species.

GENUS **SYRNIUM** SAVIGNY.

41. **Syrnium nebulosum** (FORST.).
    Barred Owl.
Resident. Not uncommon.

GENUS **MEGASCOPS** KAUP.

42. **Megascops asio** (LINN.).
    Screech Owl.
Resident. The most common of our owls. We have both the *red* and the *gray* or *mottled* forms ; the latter predominating.

GENUS **BUBO** CUVIER.

43. **Bubo virginianus** (GMEL.).
    Great Horned Owl. (Hoot Owl.)
Resident. Not common. Much less common than many years ago.

ORDER COCCYGES. CUCKOOS, ETC.

SUBORDER CUCULI. CUCKOOS, ETC.

FAMILY **CUCULIDÆ**. CUCKOOS, ANIS, ETC.

SUBFAMILY **COCCYGINÆ**. AMERICAN CUCKOOS.

GENUS **COCCYZUS** VIEILLOT.

44. **Coccyzus americanus** (LINN.).
    Yellow-billed Cuckoo.
Uncommon summer resident.

45. **Coccyzus erythrophthalmus** (WILS.).
    Black-billed Cuckoo.
Common summer resident.

SUBORDER ALCYONES. KINGFISHERS.

FAMILY **ALCEDINIDÆ**. KINGFISHERS.

GENUS **CERYLE** BOIE.

SUBGENUS **STREPTOCERYLE** BONAPARTE.

46. **Ceryle alcyon** (LINN.).
    Belted Kingfisher.
Summer resident. Not uncommon.

ORDER PICI. WOODPECKERS, WRYNECKS, ETC.

FAMILY PICIDÆ. WOODPECKERS.

GENUS DRYOBATES BOIE.

47. **Dryobates villosus** (LINN.).

Hairy Woodpecker.

Rare resident. More often seen in winter.

48. **Dryobates pubescens** (LINN.).

Downy Woodpecker.

Rather common resident, but like *D. villosus* more common in fall and winter, at which seasons it is the most common of this family.

GENUS MELANERPES SWAINSON.

SUBGENUS MELANERPES.

49. **Melanerpes erythrocephalus** (LINN.).

Red-headed Woodpecker.

Very rare in summer and fall. Has been known to breed here within a few years.

GENUS COLAPTES SWAINSON.

50. **Colaptes auratus** (LINN.).

Flicker. (Usually known here as Yellowhammer. Other local names High-hole, Wake-up, etc.)

Quite common summer resident. Has been seen here in winter.

ORDER MACROCHIRES. GOATSUCKERS, SWIFTS, ETC.

SUBORDER CAPRIMULGI. GOATSUCKERS, ETC.

FAMILY CAPRIMULGIDÆ. GOATSUCKERS, ETC.

GENUS ANTROSTOMUS GOULD.

51. **Antrostomus vociferus** (WILS.).

Whip-poor-will.

Not uncommon summer resident.

GENUS CHORDEILES SWAINSON.

52. **Cordeiles virginianus** (GMEL.).

Nighthawk.

Not uncommon summer resident.

SUBORDER CYPSELI. SWIFTS.

FAMILY **MICROPODIDÆ.** SWIFTS.

SUBFAMILY **CHÆTURINÆ.** SPINE-TAILED SWIFTS.

GENUS **CHÆTURA** STEPHENS

53. **Chætura pelagica** (LINN.).
   Chimney Swift. (Chimney Swallow.)
   Very common summer resident.

SUBORDER TROCHILI. HUMMING BIRDS.

FAMILY **TROCHILIDÆ.** HUMMING BIRDS.

GENUS **TROCHILUS** LINNÆUS.

SUBGENUS **TROCHILUS.**

54. **Trochilus colubris** LINN.
   Ruby-throated Humming Bird.
   Rather common summer resident.

ORDER PASSERES. PERCHING BIRDS.

SUBORDER CLAMATORES. SONGLESS PERCHING BIRDS.

FAMILY **TYRANNIDÆ.** TYRANT FLYCATCHERS.

GENUS **TYRANNUS** CUVIER.

55. **Tyrannus tyrannus** (LINN.).
   King-bird.
   Common summer resident.

GENUS **MYIARCHUS** CABANIS.

56. **Myiarchus crinitus** (LINN.).
   Crested Flycatcher.
   Rather rare summer resident. (I saw a pair of these birds frequently in Yale's Woods in the spring and summer of 1886, and
so late as the 15th day of June; from which fact I conclude they
were nesting there. Mr. Albert Boardman has a specimen of the
eggs of this flycatcher in his collection, which he took from a nest
in the hollow of a tree near Mr. Winfield Coe's, several years ago,
and Mr. Samuel Jackson found a nest and obtained the eggs on
Mr. Richard Miller's farm in 1884.

GENUS **SAYORNIS** BONAPARTE.

57. **Sayornis phœbe** (LATH.).
     Phœbe.
Common summer resident.

GENUS **CONTOPUS** CABANIS.

58. **Contopus virens** (LINN.).
     Wood Pewee. (Wood Phœbe.)
Common summer resident.

GENUS **EMPIDONAX** CABANIS.

59. **Empidonax minimus** BAIRD.
     Least Flycatcher.
· Summer resident.

SUBORDER OSCINES.   SONGBIRDS.

FAMILY **CORVIDÆ.**   CROWS, JAYS, MAGPIES. ETC.
SUBFAMILY **GARRULINÆ.**   MAGPIES AND JAYS.

GENUS **CYANOCITTA** STRICKLAND.

60. **Cyanocitta cristata** (LINN.).
     Blue Jay.
Not uncommon resident.   More common in fall and spring,

SUBFAMILY **CORVINÆ.**   CROWS.

GENUS **CORVUS** LINNÆUS.

61. **Corvus americanus** AUD.
     American Crow.
Resident—quite common.

FAMILY **ICTERIDÆ.**   BLACKBIRDS, ORIOLES, ETC.

GENUS **DOLICHONYX** SWAINSON.

62. **Dolichonyx oryzivorus** (LINN.).
     Bobolink.
Common summer resident.

GENUS **MOLOTHRUS** SWAINSON.

63. **Molothrus ater** (BODD.).
   Cowbird.
Common summer resident.

GENUS **AGELAIUS** VIEILLOT.

64. **Agelaius phœniceus** (LINN.).
   Red-winged Blackbird.
Common summer resident.

GENUS **STURNELLA** VIEILLOT.

65. **Sturnella magna** (LINN.).
   Meadow Lark.
Summer resident; much less common than formerly.

GENUS **ICTERUS** BRISSON.

SUBGENUS **PENDULINUS** VIEILLOT.

66. **Icterus spurius** (LINN.).
   Orchard Oriole.
Rare summer resident.

SUBGENUS **YPHANTES** VIEILLOT.

67. **Icterus galbula** (LINN.).
   Baltimore Oriole.
Common summer resident.

GENUS **SCOLECOPHAGUS** SWAINSON.

68. **Scolecophagus carolinus** (MULL.).
   Rusty Blackbird.
More or less common in spring and autumn.

GENUS **QUISCALUS** VIEILLOT.

SUBGENUS **QUISCALUS**.

69. **Quiscalus quiscula æneus** (RIDGW.).
   Bronzed Grackle. (Crow Blackbird.)
Common summer resident. Very numerous in spring and fall.
The true *Q. quiscula*, or Purple Grackle, may occur here, or at
least an intermediate form approaching it.

FAMILY **FRINGILLIDÆ**. FINCHES, SPARROWS, ETC.

Genus **PINICOLA** Vieillot.

70. Pinicola enucleator (Linn.).

  Pine Grosbeak.

Occasional winter visitant. (A few winters ago—I think during the winter of 1882–3—these birds were quite common in Meriden, visiting in small flocks the door-yards even of our city residents. They were so tame as to allow people to approach them in some instances to within a few feet.

Genus **PASSER** Brisson.

71. Passer domesticus (Linn.).

  House Sparrow. (English Sparrow.)

Abundant resident. Introduced into Meriden about twenty or twenty-one years ago by one of our prominent citizens who purchased three pairs of them in New York for which he paid eleven dollars. The blame which this man must naturally take to himself for bringing such unpopular residents into our city is doubtless lessened somewhat by the reflection that they would have reached us sooner or later anyway.

Genus **CARPODACUS** Kaup.

72. Carpodacus purpureus (Gmel.).

  Purple Finch. (Generally known in this locality as Red Linnet.)

Summer resident. Has been seen here in winter.

Genus **LOXIA** Linnæus.

73. Loxia curvirostra minor (Brehm).

  American Crossbill. (*Red Crossbill.*)

A not common winter visitor.

Genus **ACANTHIS** Bechstein.

74. Acanthis linaria (Linn.).

  Redpoll.

Irregular winter visitant.

Genus **SPINUS** Koch.

75. Spinus tristis (Linn.).

  American Goldfinch. (Yellowbird, Thistlebird.)

Common summer resident. Not uncommon through the year.

76. **Spinus pinus** (WILS.).
    Pine Siskin. (Pine Linnet.)
    Spring and autumn migrant.   Possibly winter visitant.

GENUS **PLECTROPHENAX** STEJNEGER.

77. **Plectrophenax nivalis** (LINN.).
    Snowflake. (White Snowbird.)
    More or less common winter visitant.

GENUS **POOCÆTES** BAIRD.

78. **Poocætes gramineus** (GMEL.).
    Vesper Sparrow.
    Very common summer resident.

GENUS **AMMODRAMUS** SWAINTON.

SUBGENUS **PASSERCULUS** BONAPARTE.

79. **Ammodramus sandwichensis savanna** (WILS.).
    Savanna Sparrow.
    Not uncommon summer resident.

GENUS **ZONOTRICHIA** SWAINSON.

80. **Zonotrichia albicollis** (GMEL.).
    White-throated Sparrow.
    Very common spring and autumn migrant.

GENUS **SPIZELLA** BONAPARTE.

81. **Spizella monticola** (GMEL.).
    Tree Sparrow.
    Common in fall and winter.

82. **Spizella socialis** (WILS.).
    Chipping Sparrow.
    Very common summer resident.

83. **Spizella pusilla** (WILS.).
    Field Sparrow.
    Common Summer resident.

GENUS **JUNCO** WAGLER.

84. **Junco hyemalis** (LINN.).

Slate-colored Junco. (Black Snow Bird.)

Very common in the fall.   More or less common in winter and spring.

GENUS **MELOSPIZA** BAIRD.

85. **Melospiza fasciata** (GMEL.).

Song Sparrow.

Very common summer resident.

GENUS **PASSERELLA** SWAINSON.

86. **Passerella iliaca** (MERR.).

Fox Sparrow.

Spring and fall migrant.

GENUS **PIPILO** VIEILLOT.

87. **Pipilo erythrophthalmus** (LINN.).

Towhee. (Ground Robin, Chewink or Pewink.)

Common summer resident ; often staying till quite late in the fall.

GENUS **HABIA** REICHENBACH.

88. **Habia ludoviciana** (LINN.).

Rose-breasted Grosbeak.

Summer resident ; formerly rare—more common late years.

GENUS **PASSERINA** VIEILLOT.

89. **Passerina cyanea** (LINN.).

Indigo Bunting.

Rather common summer resident.

FAMILY **TANAGRIDÆ** TANAGERS.

GENUS **PIRANGA** VIEILLOT.

90. **Piranga erythromelas** VIEILL.

Scarlet Tanager.

Summer resident.   Not very common.

FAMILY **HIRUNDINIDÆ.** SWALLOWS.

Genus **PROGNE** Boie.

91. **Progne subis** (Linn.).

Purple Martin.

Summer resident; formerly quite common; more rare of late years.

Genus **PETROCHELIDON** Cabanis.

92. **Petrochelidon lunifrons** (Say).

Cliff Swallow. (Eave Swallow.)

Summer resident. Not common, but builds regularly in some parts of the town.

Genus **CHELIDON** Forster.

93. **Chelidon erythrogaster** (Bodd.).

Barn Swallow.

Very common summer resident.

Genus **TACHYCINETA** Cabanis.

94 **Tachycineta bicolor** (Vieill.).

Tree Swallow.

Uncommon summer resident.

Genus **CLIVICOLA** Forster.

95. **Clivicola riparia** (Linn.).

Bank Swallow. (Sand Martin.)

Not uncommon summer resident. (A few build every year in a bank near the Foster Brothers' Pond.)

Family **AMPELIDÆ.** Waxwings, Etc.

Subfamily **AMPELINÆ.** Waxwings.

Genus **AMPELIS** Linnæus.

96 **Ampelis cedrorum** (Vieill.).

Cedar Waxwing. (Cedar Bird.)

Not very common summer resident. More numerous in fall and spring. and seen occasionally in sheltered localities during the winter.

Family **LANIIDÆ.** Shrikes.

Genus **LANIUS** Linnæus.

97. Lanius borealis Vieill..

Northern Shrike. (Butcher Bird.)

Occasional winter visitant. (I have noticed that some egg collectors in this vicinity are calling the Crested Flycatcher the Butcher Bird.)

Family **VIREONIDÆ.** Vireos.

Genus **VIREO** Vieillot.

Subgenus **VIREOSYLVA** Bonaparte.

98. Vireo olivaceus (Linn.).

Red-eyed Vireo.

Very common summer resident.

99. Vireo gilvus (Vieill.).

Warbling Vireo.

Rather common summer resident.

Subgenus **LANIVIREO** Baird.

100. Vireo flavifrons Vieill.

Yellow-throated Vireo.

Rather common summer resident.

101. Vireo solitarius (Wils.)

Blue-headed Vireo.

Spring and autumn migrant.   Not common.

Subgenus **VIREO** Vieillot.

102. Vireo noveboracensis (Gmel.)

White-eyed Vireo.

Rather common summer resident.

Family **MNIOTILTIDÆ.** Wood-Warblers.

Genus **MNIOTILTA** Vieillot.

103. Mniotilta varia (Linn.).

Black and White Warbler. (Black and White Creeping Warbler.)

Very common summer resident.

Genus HELMINTHOPHILA Ridgway.

104. **Helminthophila ruficapilla** (Wils.).

Nashville Warbler.

Common summer resident.

Genus COMPSOTHLYPIS Cabanis.

105. **Compsothlypis americana** (Linn.).

Parula Warbler. (Blue Yellow-backed Warbler.)

Probably rare summer resident. I have never seen it except during the spring migrations no later than May 11, and then only rarely. Mr. H. H. Kendrick found an old nest last summer which in my opinion was a nest of this species of warbler.

Genus DENDROICA Gray.

Subgenus PERISSOGLOSSA Baird.

106. **Dendroica tigrina** (Gmel.)

Cape May Warbler.

Rare spring and autumn migrant. I have never seen but one bird of this species in Meriden (Yale's Woods, May 26th, 1885).

Subgenus DENDROICA Gray.

107. **Dendroica æstiva** (Gmel.).

Yellow Warbler. (Summer Yellow Bird.)

Very common summer resident.

108. **Dendroica cærulescens** (Gmel.).

Black-throated Blue Warbler.

Only during the migrations. Rather rare.

109. **Dendroica coronata** (Linn.).

Myrtle Warbler. (Yellow-rumped Warbler.)

Only during the migrations.

110. **Dendroica maculosa** (Gmel.).

Magnolia Warbler. (Black and Yellow Warbler.)

Only during the migrations.

111. **Dendroica pennsylvanica** (Linn.).

Chestnut-sided Warbler.

Rare summer resident. Not often observed even during the migrations.

112.   **Dendroica striata** (Forst.).
Black-poll Warbler.
Spring and autumn migrant.

113.   **Dendroica blackburniæ** (Gmel.).
Blackburnian Warbler.
Only spring and autumn migrant.

114.   **Dendroica virens** (Gmel.).
Black-throated Green Warbler.
Spring and fall migrant.

115.   **Dendroica palmarum hypochrysea** (Ridgw.).
Yellow Palm Warbler. (Yellow Red-poll.)
More or less common during the migrations.

Genus **SEIURUS** Swainson.

116.   **Seiurus aurocapillus** (Linn.).
Oven-bird.
Quite common summer resident.

117.   **Seiurus noveboracensis** (Gmel.).
Water Thrush.
Spring and fall migrant.   I saw a pair of these birds in a swamp on the Greer farm on the first day of June, 1884, which was so late in the season that I presume they stayed through the summer.

Genus **GEOTHLYPIS** Cabanis.

Subgenus **GEOTHLYPIS** Cabanis.

118.   **Geothlypis trichas** (Linn.).
Maryland Yellow-throat.
Quite common summer resident.

Genus **ICTERIA** Vieillot.

119.   **Icteria virens** (Linn.).
Yellow-breasted Chat.
Occasional summer resident.   Commonly seen only in the spring.

Genus **SETOPHAGA** Swainson.

120. **Setophaga ruticilla** (LINN.).
   American Redstart.
Very common summer resident.

FAMILY **TROGLODYTIDÆ.** WRENS, THRASHERS, ETC.

SUBFAMILY **MIMINÆ.** THRASHERS.

GENUS **MIMUS** BOIE.

121. **Mimus polyglottos** (LINN.).
   Mockingbird.
Extremely rare summer visitant.   The only authentic instance I
have of the taking of this bird in Meriden dates back about thirty
years ago, when one was shot in the southern part of the town.

GENUS **GALEOSCOPTES** CABANIS.

122. **Galeoscoptes carolinensis** (LINN.).
   Catbird.
Very common summer resident.

GENUS **Harporhynchus** CABANIS.

SUBGENUS **METHRIOPTERUS** REICHENBACH.

123. **Harporhynchus rufus** (LINN.).
   Brown Thrasher.
Very common summer resident.

SUBFAMILY **TROGLODYTINÆ.** WRENS.

GENUS **TROGLODYTES** VIEILLOT.

SUBGENUS **TROGLODYTES.**

124. **Troglodytes aëdon** VIEILL.
   House Wren.
More or less common summer resident.   Not as plenty as for-
merly.

FAMILY **CERTHIIDÆ.** CREEPERS.

GENUS **CERTHIA** LINNÆUS.

125. **Certhia familiaris americana** (BONAP.).
   Brown Creeper.
Rather rarely seen in autumn or winter.

FAMILY **PARIDÆ**. NUTHATCHES AND TITS.

SUBFAMILY **SITTINÆ**. NUTHATCHES.

GENUS **SITTA** LINNÆUS.

126. Sitta carolinensis (LATH.).
White-breasted Nuthatch.

Rather rare resident. More common in spring, autumn and winter.

SUBFAMILY **PARINÆ**. TITMICE.

GENUS **PARUS** LINNÆUS.

SUBGENUS **PARUS** LINNÆUS.

127. Parus atricapillus LINN.
Chicadee.

Resident; but not common in the summer. Quite common in fall and winter.

FAMILY **SYLVIIDÆ**. WARBLERS, KINGLETS, GNATCATCHERS.

SUBFAMILY **REGULINÆ**. KINGLETS.

GENUS **REGULUS** CUVIER.

128. Regulus satrapa LICHT.
Golden-crowned Kinglet.

Only during the winter and spring, and then only occasional.

129. Regulus calendula (LINN.).
Ruby-crowned Kinglet.

Spring and fall migrant.

FAMILY **TURDIDÆ**. THRUSHES, SOLITAIRES, STONECHATS, BLUEBIRDS, ETC.

SUBFAMILY **TURDINÆ**. THRUSHES.

GENUS **TURDUS** LINNÆUS.

SUBGENUS **HYLOCICHLA** BAIRD.

130. Turdus mustelinus GMEL.
Wood Thrush.

Very common summer resident.

131. **Turdus fuscescens** STEPH.
    Wilson's Thrush. (Veery.)
  Common summer resident.

132. **Turdus aliciæ** BAIRD.
    Gray-cheeked Thrush.
  Only during the migrations.

133. **Turdus ustulatus swainsonii** (CAB.).
    Olive-backed Thrush.
  Only during the migrations.

134. **Turdus aonalaschkæ pallasii** (CAB.).
    Hermit Thrush.
  Spring and autumn migrant. Not often seen or else mistaken
  for Wood Thrush or for Wilson's Thrush.

GENUS **MERULA** LEACH.

135. **Merula migratoria** (LINN.).
    American Robin.
  Very common summer resident.   Occasional winter resident.

GENUS **SIALIA** SWAINSON.

316. **Sialia sialis** (LINN.).
    Bluebird.
  Common summer resident.

Additional Plants

Found growing in Meriden, Conn., since issue of Catalogue in 1885.

IDENTIFIED BY MRS. E. B. KENDRICK.

(*Director of the Botanical Department of the Meriden Scientific Association.*)

Order 1. RANUNCULACEÆ.
RANUNCULUS, L.

fascicularis, Muhl. Buttercup.

Order 4. BERBERIDACEÆ.
PODOPHYLLUM, L.

peltatum, L. Mandrake.

Order 9. CRUCIFERÆ.
DENTARIA, L.

laciniata, Muhl. Pepper-root.

Order 10. VIOLACEÆ.
VIOLA, L.

pubescens, Ait. var. **scabriuscula,** Torr. and Gray.

Order 14. CARYOPHYLLACEÆ.
SILENE, L.

INFLATA, Smith. Bladder Campion.

ANYCHIA, Michx.

dichotoma, Michx. Forked Chickweed.

Order 15. PORTULACACEÆ.

PORTULACA, Tourn.

GRANDIFLORA, Gray. Purslane. Escaped.

Order 19. GERANIACEÆ.

GERANIUM, L.

PUSILLUM, L. Small-flowered Cranesbill.

ERODIUM, L'Her.

CICUTARIUM, L'Her. Storksbill.

Order 27. LEGUMINOSÆ.

CORONILLA, L.

varia, L. Purple Coronilla. Escaped.

Order 28. ROSACEÆ.

POTENTILLA, L.

Canadensis, L. var. simplex, Torr. and Gray. Cinque-foil.

AMELANCHIER, Medic.

Canadensis, Torr. and Gray. var. oblongifolia. Shad-bush.

Order 36. UMBELLIFERÆ.

HERACLEUM, L.

lanatum, Michx. Cow-Parsnip.

Order 39. CAPRIFOLIACEÆ.

SAMBUCUS, Tourn.

pubens, Michx. Red-berried Elder.

Order 42. COMPOSITÆ.

DIPLOPAPPUS, Cass.

linariifolius, Hook. Double-bristled Aster.

Order 45. ERICACEÆ.

VACCINIUM, L.

Pennsylvanicum, Lam. Dwarf Blueberry.

Order 51. OROBANCHACEÆ.

EPIPHEGUS, Nutt.

Virginiana, Bart. Beechdrops.

Order 52. SCROPHULARIACEÆ.
VERONICA, L.
Virginica, L. Culver's Root.

GERARDIA, L.
pedicularia, L. Small False Foxglove.

ILYSANTHES, Raf.
gratioloides, Benth. False Pimpernel.

Order 54. VERBENACEÆ.
PHRYMA, L.
Leptostachya, L. Lopseed, Rare.

Order 62. ASCLEPIADACEÆ.
ASCLEPIAS, L.
obtusifolia, Michx. Milkweed.

Order 74. JUGLANDACEÆ.
CARYA, Nutt.
tomentosa, Nutt. White-heart Hickory.

Order 75. CUPULIFERÆ.
QUERCUS, L.
bicolor, Willd. Swamp White Oak.
coccinea, Wang. Scarlet Oak.

CORYLUS, Tourn.
Americana, Walt. Wild Hazel-Nut.

Order 77. BETULACEÆ.
BETULA, Tourn.
lutea, Michx, f. Yellow Birch.

Order 78. SALICACEÆ.
SALIX, Tourn.
myrtilloides, L. Myrtle-leaved Willow.
DECIPIENS, Hoffm. Brittle Willow.

POPULUS, Tourn.
DILATATA, Ait. Lombardy Poplar.

Order 79. CONIFERÆ.

LARIX, Tourn.

Americana, Michx. Tamarack.

JUNIPERUS, L.

communis, L. Juniper.

Virginiana, L. Red Cedar.

Order 87. LILIACEÆ.

ASPARAGUS, L.

OFFICINALIS, L. Garden Asparagus. Escaped.

Order 92. CYPERACEÆ.

ERIOPHORUM, L.

polystachyon, L. Cotton-grass.

CAREX, L.

miliacea, Muhl. Sedge.

Order 93. GRAMINEÆ.

GYMNOSTICHUM, Schreb.

Hystrix, Schreb. Bottle-brush Grass.

Order 95. FILICES.

CAMPTOSORUS, Link.

rhizophyllus, Link. Walking Leaf. Rare.

CYSTOPTERIS, Bernhardi.

fragilis, Bernh. var. DENTATA, Hook. Bladder Fern.

PHEGOPTERIS, Fée.

hexagonoptera, Fée. Beech Fern.

Order 96. LYCOPODIACEÆ.

LYCOPODIUM, L., Spring.

complanatum, L. Ground Pine.

West Peak, and What It Saith.

By Rev. J. T. Pettee.

[West Peak is the name which we, Meridenites, give to the most westerly of our " Hanging Hills." It is, by Prof. Guyot's survey, 995 feet above the waters of the Sound, and, though far from being the highest mountain in the State (Mt. Brace in Saulsbury being 2225 feet high), is, by considerable, the highest of the trap dikes of the Connecticut Valley. Geologists are agreed, I believe, in thinking that the valley, which stretches from Hartford to New Haven, was once an estuary or arm of the sea, and Percival, the distinguished geologist of Connecticut, was the first to show how, by the eruption of the trap across the valley in Meriden, the Connecticut River was made to change its course, and empty at Saybrook instead of New Haven. By a poetic license which, I think, perfectly pardonable, I have taken a part for the whole, and spoken of West Peak as being formed under the ocean.

That the trap was erupted under water seems to be the concurrent opinion of geologists from Lyell to Le Conte. Says Lyell (Elements of Geology, p. 115), " The recent examiation of the igneous rocks of Sicily has proved that all the more ordinary varieties of European trap have been there produced *under the waters of the sea.*" And Le Conte, in his admirable " Elements," p. 206, says "Sometimes similar sheets (of trap) are found between the strata (of sandstone,) as if *outpoured on the sea-bottom,* and afterwards covered with sediment." And Dana everywhere speaks of the trap of the Connecticut valley as erupted *through the sandstone,* which was, of course, formed under water.

As Owen says that the *mastodon* ranged " throughout the tropical and temperate latitudes," and as its bones have been found (in this case associated with a human skull,) as near as Worcester, Mass., on the east, and Cohoes and Newburg, N. Y., on the west; and as Dr. Percival (Geological Report of Conn., p. 465,) says " a vertebra of a mammoth was found in New Britain," I think it no stretch of the imagination to conceive of the mastodons as " roaming through these lands ;" especially in view of the thought that the mammoth spoken of by Dr. Percival was probably a mastodon, the difference between the vertebrae of these mammals being hardly distinguishable.

The admirable article by Prof. T. Stery Hunt, Mass- Inst. Tech., Boston, *Fossil Footprints,* in the seventh Vol. of Appleton's American Cyclopaedia, is my authority for my prehistoric animals. " More recent studies of the fossil remains of these animals, which have been carefully made by various naturalists, and especially by Cope, have made us acquainted with that curious class of animals, the *dinosaurs.* These creatures constituted numerous genera and species, some of gigantic size, others comparatively small ; some feeding on plants and others carniverous ; but all remarkable for presenting a higher tyye of reptilian organizaticn than any now existing, and approaching in some respects to the birds and in others to the mammalia. Among the vegetable feeders of this group was the *hadrosaurus*, a gigantic animal, 20 feet or more in height, with huge bird-like legs and feet, a lizard like tail, a diminutive head, and small fore feet or hands, feeding on plants ; while *læleps* was an equally large carniverous animal of somewhat similar organization."—J. T. P.]

For ages, when the world was young,
　　I slept upon my lava-bed,
While sandstones formed, and oceans sung
　　Their solemn anthems o'er my head.

Ages on ages rolled away,
　　The wrinkled earth itself grew old ;
And still upon my bed I lay,
　　Oppressed by weight and years untold.

The ocean still above me rolled,
　　The sandstone strata thicker grew ;
I lay and groaned beneath the Old,
　　Crushed and encumbered by the New.

Then in a glad auspicious hour,
　　Which made my rocky heart rejoice.
I felt a resurrection power—
　　I heard a resurrection voice.

It said " O mountain, 'wake, arise ;
　　Throw off the sandstone from thy breast ;
Roll back the seas, and 'neath the skies
　　Show the bold frontage of thy crest."

I woke as from a troubled dream ;
　　Threw off the weight by power divine ;
Rose to the sun's refulgent beam,
　　And stellar orbs that round me shine.

The frighted waters sought the sea ;
　　The rifted sandstone opened wide,
And I, aglow with light of day,
　　Rejoiced, a Mountain in my pride.

Nor I alone : On every hand
　　Around my peak like mountains stand,
Which heard the voice, and felt the power,
　　That raised me in my natal hour.

South Mountain, Cat Hole, by my side,
 Almost as bold and steep as I,
Majestic in their mountain pride,
 Point their tall turrets to the sky.

High Rock and Rattlesnake arise ;
 Newgate and Talcott farther on ;
And resting on the northern skies,
 Proud peaks of Holyoke and Mt. Tom.

Northeast Mt. Lamentation stands,
 Higby and Besec, Middletown,
To Durham ranges stretch their hands,
 Where Tremont towereth all their own.

Totoket rises farther down ;
 And Pistapaug and Saltonstall
Raise to the skies their walls of stone,
 Their mural castles gaunt and tall.

Near on my south Mt. Carmel lies,
 A giant slumbering in his might ;
East Rock and West Rock kiss the skies,
 And Whitney Peak delights the sight.

While on my West, in peaks less bold,
 The same Plutonian power is seen,
Trappean dikes of lava cold,
 And sandstone tilted thrown between.

These lesser heights, whose waving lines
 Such beauty to the landscape give,
Tell of the old Triassic times,
 And to my tale their witness give.

The voice which called *me* from the deep
 These trappean mountains all did hear,
And rose with me from natures sleep,
 And stand, as I stand, proudly here.

And now for long telluric years,
 I've stood a sentry o'er the land,
And watched with varying hopes and fears,
 The changes of Time's mighty hand.

I saw the glacier in his might
 Sweep from the north, a frozen sea,

Ice piled on ice to mountain height,
  Moving, methought, resistlessly.

I felt his cruel ice-bound teeth
  Plough in my flanks, as on his way
He ground and crushed my rocks beneath :
  I show the furrows to this day.

I've seen the new-formed earth press back
  The waters of the refluent sea,
And wide diffused along its track
  The germs of floras yet to be.

I've seen the living germs far borne
  By wind and wave from other lands,
Grasses and grains, and trees uptorn,
  Find lodgement in the new-made sands.

And as these sands have richer grown,
  And turned to mould from year to year,
From seeds by wind and water sown
  The flowerets and the plants appear.

I saw the primal forests rise,
  And slowly grow from year to year,
Until their branches swept the skies,
  And veiled the landscape far and near.

I saw the beasts and birds appear ;
  Beheld their forms and heard their song,
Prowling or flying, without fear
  These mountains and these hills along.

The *mastodon*, the mammal's pride,
  Unchallenged roamed through all these lands,
And birds, immense in height and stride,
  Left their deep foot-prints on the sands.

Huge *dinosaurs* and *hadrosaurs*
  Stalked through this vale with awful form ;
Lœleps, with dread carniverous jaws,
  Defied the prehistoric storm.

These on the land ; while, in the sea,
  And every stream that marked this vale,
The fishes played in sportive glee,
  And left their impress on the shale.

Flora and Fauna of that age :
 Your fossils, treasured in my rocks,
The wisdom of the world engage
 To find the key these stores unlocks.

I saw the prehistoric race
 Roam o'er these vales, and climb these hills ;
The mountain caves their dwelling place,
 Their drinking place the mountain rills.

Their implements of rudest stone,
 Alike of industry or war ;
Their needles formed of pointed bone,
 Their garments of the wild beast's fur.

For centuries they held these lands,
 But left no mounds to mark their day,
Dissolved, like snow, their scattered bands;
 They lived, and died, and passed away.

I saw the red man when he came,
 And watched him here for many a year;
Leaving behind him but a name,
 At last I saw *him* disappear.

I saw the lordly white man come,
 And take possession of this vale;
These hills and valleys call his own ;
 And here begins my modern tale.

I've seen two centuries come and go,
 Yea, nearly three, since first was prest
The Indian's harbinger of woe,
 The white man's foot, upon my breast.

I've seen the change by centuries wrought
 Engraved in Progress' deepest lines ;
To us with greater interest fraught
 Than those of old Triassic times.

I've seen the wilderness subdued,
 Fair villages and towns arise ;
Cities with energy imbued,
 And art and skill and bold emprise.

A hundred such around me rise ;
  I see them from my mountain height ;
Their gilded domes and cloud capped spires
  Lend fair enchantment to the sight.

Their business gongs salute my ear,
  Their throbbing engines jar my crest ;
Their mighty industries appear,
  Which meet no check, and know no rest.

Of all the the towns that round me rise,
  Of all the cities that I greet,
There's none seems fairer to my eyes
  Than that which slumbers at my feet.

Fair city of the Silver Art,
  Still slumber in thy quiet vale ;
With rocky fastnesses begirt,
  May naught against thy peace prevail.

Long will I guard thy schools and homes,
  And hold thy precious interests fast,
Watching thy good for years to come,
  As I have watched it in the past.

Now, lest my boast I should renew
  Of what I've seen and what I've done,
I know my friends, as well as you,
  Naught can endure beneath the sun.

If standing in my mountain strength,
  I've seen the earth with ruin strewn,
I must confess the truth at length—
  I stand a witness of my own.

My peak fast crumbling year by year.
  The debris gathering round my feet,
Proclaim in accents loud and clear.
  The doom which men and mountains meet.

This doom keeps sounding in my ears,
  And rolls along this mountain range,
Nothing remains unchanging here—
  But change itself, all else is change.

Ye who oft climb my summit fair,
  And clamber o'er my fallen stones,
Yourselves for every change prepare—
  Ye climb my summit o'er my bones.

Mountains and mortals must decay ;
  Crumble alike like grains of sand ,
And when all else has passed away,
  The Mount of Truth alone will stand.

[For the *geology* of West Peak see Prof. James D. Dana's " Excursion to the Hanging Hills of Meriden," as given in Davis' History of Meriden, pp. 53–66.]

[For fossils of Connecticut Valley, see Art. Fossil Foot-prints in Appleton's Am. Encyclopedia.]

OF THE

ᛗERIDEN SCIENTIFIC ·ASSOCIATION·

MERIDEN, CONN.

VOL. III.

1887–1888.

PROCEEDINGS AND TRANSACTIONS

OF THE

SCIENTIFIC ASSOCIATION,

MERIDEN, CONN.

1887–1888.

Vol. III.

MERIDEN, CONN.
E. A. HORTON & CO., PRINTERS.
1889.

CONTENTS.

PUBLISHING COMMITTEE.

CHAS. H. S. DAVIS, M D.   FRANKLIN PLATT.

MRS. E. B. KENDRICK.

Report of Secretary.

At the beginning of the ninth year of the Association we have on our books the names of some one hundred and fifty members.

During 1887 nine gentlemen and five ladies, and, in 1888, eight gentlemen and one lady were elected members of the Association.

The meetings of the Association are generally well attended, and at each alternate month the Directors of the various sections have presented a resumé of scientfic progress made during the previous two months.

At each alternate month the evening is devoted to one or two papers by members of the Association and others.

The following papers were read before the Association in 1887:

Some Dictionary Errors, . .	GEO. L. COOPER.
Some New Theories in Regard to Water, .	F. T. IVES.
Percival as a Geologist. . . .	REV. J. T. PETTEE.
Review of Sir John Lubbock's "Ants, Wasps and Bees," .	. REV. S. H. WATKINS.
Language of India,	REV. GEO. H. McGREW.
Japan, . .	CHAS. M. CRAWFORD.
Oceanic Islands,	MRS. A. A. FRENCH.

The following papers were read before the Association in 1888:

Talk on Electricity, ...	W. G. RIGGS.
Review of Prof. Powell's " Lectures on Human Progress," .	. HENRY S. PRATT.
Appeal to Thinkers,	CLARENCE E. ELLSBREE.
Elocution as Taught at Monroe College.	. SUSIE D. DREW.

Diamond Fields of Africa, .	HEMAN B. ALLEN.
Animal Effigies of Wisconsin,	WEGIA HOPE HALL.
Geological Formation of Asia,	REV. J. H. CHAPIN.
Benefits, Dental and Otherwise,	CHAS. C. BARKER.

In July, 1887, the members of the Association, with friends, made an excursion to the " ash bed " in the Berlin woods.

In 1888 a new case was purchased for books and specimens.

Through the kindness of the Common Council their room is still used for the meetings of the Association, but it is hoped that soon we can have rooms of our own, in order that we can make a proper display of our collections.

CHAS. H. S. DAVIS,

Secretary.

Report of Treasurer.

MERIDEN SCIENTIFIC ASSOCIATION.

1887.

Dr.

January 1.

To Paid Dr. Davis,	.	$4.00
" E. L. Marvin,		2.00
" F. Platt, . .		6.50
" E. A. Horton & Co., .		98.30
" H. Wales Lines & Co.,		36.00
Balance,		205.40
		$352.20

Cr.

By Balance,		$157.52
" Dues, etc.,	. . .	194.68
		$352.20

1888.

Dr.

January 1.

To Balance, . .	$205.40
_ " Received from Collector, .	143.00
" " Interest on Loans, .	5.14
	$353.54

Cr.

By Paid Mrs. E. B. Kendrick,	$10.80	
" H. Wales Lines Co., .	.80	
" J. T. Pettee (Hall Rent),	5.00	
" C. L. Little, . .	53.53	
" Peck Bros.,	1.90	
" E. P. Judd,	24.80	
" W. M. Quested,	1.50	
" Eaton & Peck Co.,	5.00	
" H. N. Brooks & Co..	2.00	
" E. A. Horton & Co.,	13.25	
" C. L. Little,	.35	
		118.93

Balance on hand, . $234.61

CLARENCE E. ELLSBREE,

Treasurer.

January 1, 1889.

Report of Curator and Librarian.

I have to report the following additions to the Library and Cabinet of the Association during the years 1887 and 1888:

Peabody Academy of Science. Nineteenth Annual Report.

Transactions of the Connecticut Academy of Arts and Sciences. Vol. 7. Parts 1 and 2.

Bulletins of the California Academy of Arts and Sciences. January, June and November, 1887.

Transactions of the Academy of Science of St. Louis, 1878-1886.

Bulletin of the Torrey Botanical Club. 1887-1888.

American Museum of Natural History of New York City. Reports for 1886, 1887 and 1888.

American Association for the Advancement of Science. Annual Report for 1888.

Notes on Juessia, by Chas. Wright.

History and Work of the Warner Observatory, Rochester, for 1883-1886.

Memoir of Rev. Elisha Mitchell.

Journal of the Elisha Mitchell Scientific Society. 1885-1888.

Proceedings of National Scientific Association. 1887-1888.

Transactions of Vassar Bros.' Institute, Scientific Section. Vol. 4.

Journal of Comparative Medicine and Surgery, April and October, 1888.

Transactions of Nova Scotian Institute of Natural Science. 1884-1887.

Seventh Annual Report of State Mineralogist of California.

Proceedings of the Royal Dublin Scientific Society, 1888.

West American Scientist, 1887-1888.

Proceedings and Transactions of Natural History Society of Glasgow. Vol. 1. Part 3.

Brookville Society of Natural History. Bulletin No. 2.

Report of Superintendent of Nautical Almanac, 1887.

Man and Nature. G. P. Marsh. 1869.

The Races of Man. O. Peschel. 1876.

Report on Cotton Insects. J. H. Comstock. 1879.

The Geology and Veins of Tombstone, Arizona. W. P. Blake. 1881.

Silver Mining and Milling at Butte, Montana. W. P. Blake. 1887.

Nickel. W. P. Blake. 1885.

Description of a Meteorite from Tenn. W. P. Blake. 1886.

Discovery of Tin Ore in the Black Hills, Dak. W. P. Blake. 1883.

Tin. W. P. Blake. 1885.

Antimony. W. P. Blake. 1885.

Report on Iron and Steel. W. P. Blake. 1876.

Catalogues of Amherst College. 1886, 1887, 1888.

XVII Annual Report of the Curators of the Museum of Weslyan University.

Proceedings of the Boston Scientific Society. 1887.

Proceedings of the Natural Science Association of Staten Island. 1888.

Characters of Some New Musci. C. Wright.

Biology. Huxley.

Butterflies of New England. Maynard.

Manual of the Vertebrates. D. S. Jordan. 1884.

Zoology. A. S. Packard. 1886.

Key to North American Birds. E. Coves. 1872.

Description of the Insects of North America. T. Say. 2 Vols.

Report of U. S. Geological Survey. Vol. 3. Tertiary Vertebrata. 1884.

American Ephemeris. 1887-88-89-90-91.

Report on the Geology of the High Plateaus of Utah. C. E. Dutton. 1880.

Tribes of the Extreme Northwest. W. H. Dall. 1877.

Tribes of Western Washington and Northwestern Oregon. G. Gibbs. 1877.

Houses and House-life of the American Aborigines. L. H. Morgan. 1881.

The Fishery Industries of the U. S. C. B. Goode. 2 Vols. 1884.

United States Geological Survey. J. W. Powell. 1884-85.

The Forms of Water in Clouds and Rivers, Ice and Glaciers. J. Tyndall. 1876.

The Five Senses of Man. J. Bernstein. 1876.

Animal Mechanism. E. J. Marey. 1879.

The Correlation and Conservation of Forces. E. L. Youmans. 1872.

The Ancient World. D. T. Ansted. 1847.

Abhandlungen Herausgegeben vom Naturwissenschaftlichen Vereine zu Bremen. X Bd. 1-2 Heft. 1888.

Jahresbericht der Naturhistorischen Gesellschaft zu Hanover. 1883-87.

Neunundzwanzigster Bericht der Naturwissenschaftlichen fur Schwaben und Neuberg in Augsburg. 1887.

Publications de L'Institute Royal Grand Ducal de Luxembourg. Tome XX. 1886.

Archivos do Museu Nacional do Rio de Janeiro. Vol. 7. 1887.

Observations Meteorologiques faites a Luxembourg, moyennes de la Périod de 1854-1883. 1887.

SPECIMENS.

Copper Calcite, Mississippi.

Lead Carbonate, Dakota.

Zinc, Pennsylvania.

Geodes of Calamine, Michigan.

Manganese, Michigan.

Galena, Dakota.

Crystalized Quartz, Minnesota.

Copper in Conglomerate, Michigan.

Moss Copper in Calcite, Michigan.

Magnetic Ore, Michigan.

Franklinite, New Jersey.

Iron Ore, Michigan.

Manganese, Mississippi.

Copper, Michigan.

Gold-bearing Quartz and Pyrites, Michigan.

Fossil Fish, Rock Falls

Block of Graphite, Pyrites and Asbestos, Michigan.

Petrified Wood, Michigan.

Geode of Quartz, Jackson, Mississippi.

Fossil Plant, Michigan.

Buhrstone, Paris, France.

Calcarious Tufa, Syracuse, N. Y.

Porphery, Deadwood, Dakota.
Hematite and Jasper, Jackson, Miss.
Crystaline Pyrites, Michigan.
Hornblende, Michigan.
Copper, Calcite and Amygaloid, Michigan.
Black Graphite, Michigan.
Graphite, Dakota.
Rock Fern, Michigan.
Hot Springs Tufa, Arkansas.
Black Hills Gypsum, Deadwood, Dakota.
Dendritus Limestone, Dakota.
Specular Hematite Iron, Michigan.
Ash Rock, Reservoir, Meriden.
Horned Toad, Texas.
Black Snake, Meriden.
Newt, Meriden.
Flying Fish, California.
Sea Porcupine, California.
Sea Egg, California.
Star Fish, California.
Echinus, California.
Fan Coral, California.
Petrified Wood, Oregon.
Indian Arrow Heads, Oregon.
Hawksbill Turtle, Magdalen.

ROBERT BOWMAN,

Curator and Librarian.

January, 1889.

PROCEEDINGS.

1887.

Regular meeting of the Association, January 11, 1887. Dr. O. D. J. Hughes and Samuel Jackson were proposed for membership.

Mrs. McGrew, Mrs. Mary Lowell, Dr. A. W. Tracy and Mr. and Mrs. Geo. L. Hall were elected members of the Association.

Voted, that the collector be allowed ten per cent. for collecting the dues of the Association.

Voted, that Mrs. Ella B. Kendrick act as collector for the ensuing year.

Dr. G. H. Wilson was appointed Director of the department of Microscopy; Wm. H. Doyle of Entomology; Geo. W. Smith of Necrology; Mrs. C. E. Ellsbree of Geography, and F. E Sands of Engineering.

Voted, that a committee be appointed by the President to procure an Act of Incorporation. Messrs. F. Platt, W. W. Lee and G. H. Wilson were appointed as the committee.

Voted, that the President appoint a committee to look for rooms for the Association. Dr. Davis, C. E. Ellsbree and R. Bowman were appointed as the committee.

The following officers were elected for the ensuing year:

President—REV. J. H. CHAPIN.

Vice-President—REV. J. T. PETTEE.

Secretary—DR. C. H. S. DAVIS.

Treasurer—A. B. MATHER.

Curator and Librarian—ROBERT BOWMAN.

Adjourned to meet Monday evening, February 14, 1887.

FEBRUARY, 14, 1887.

Regular meeting of the Association, Monday evening, February 14, 1887. Dr. O. D. J. Hughes and Samuel Jackson were elected members of this Association.

Wm. G. Hooker was proposed for membership.

Voted, that the bill of Messrs. Davis and Marvin be paid.

H. B. Allen, S. J. Robie and Mrs. Ellen Cooper were appointed committee on programme for the April meeting.

Voted, that it shall be the duty of the presiding officer at any meeting of this Association, to read any question of scientific interest which shall have been received from any member since the last meeting, and to assign such question to the director of the proper section, or to any member for answer.

Adjourned to meet Monday evening, March 14, 1887.

MARCH 14, 1887.

Regular Meeting of the Association Monday evening, March 14, 1887.

Wm. G. Hooker was elected a member of the Association.

Mrs. Mary E. Rogers, Wm. E. Gard, Mrs. Mary A. Gard and Joseph Girard were proposed for membership.

Committee on lectures, appointed at the meeting of the Association in May, reported, and the committee was continued, with the addition of the President and Vice-President.

Voted, that the Act of Incorporation be accepted.

Committee on room reported, and the committee was continued.

Adjourned to meet Monday evening, April 11, 1887.

APRIL 11, 1887.

Regular meeting of the Association Monday evening, April 11, 1887.

Mrs. Mary E. Rogers, Mr. and Mrs. Wm. E. Gard and Joseph Girard were elected members of the Association.

Mrs. A. A. French, Mrs. Nelson H. Ives and H. C. Venter were proposed for membership.

Report of lecture committee was accepted, and committee was discharged.

Committee on rooms reported, and the committee was continued.

R. Bowman, W. W. Lee and Mrs. F. Platt were appointed committee on programme for the June meeting.

Voted, that the Secretary notify the members of the regular

meetings of the Association, by means of postal card, for the May, June and September meetings.

Dr. Chapin, G. L. Cooper and Mrs. Kendrick were appointed committee on excursion.

Adjourned to meet Monday evening, May 9, 1887.

MAY 9, 1887.

Regular meeting of the Association Monday evening, May 9, 1887.

Mrs. A. A. French, Mrs. N. H. Ives and Mr. H. C. Venter were elected members of the Association.

Committee on programme for the June meeting reported, and the committee was continued.

Committee on excursion reported, and the committee was continued with instructions to make arrangements for an excursion to Glastonbury.

Committee on rooms reported, and the committee continued, to make a report at the June meeting.

Adjourned to meet Monday evening, June 13, 1887.

JUNE 13, 1887.

Regular meeting of the Association Monday evening, June 13, 1887.

Committee on programme reported, and the committee was discharged.

Committee on excursion reported, and the committe was continued.

Voted, that the Association make an excursion to Glastonbury, June 25.

Adjourned to meet Monday evening, September 12, 1887.

SEPTEMBER 12, 1887.

Regular meeting of the Association Monday evening, September 12, 1887.

Mrs. K. A. Hathaway was proposed for membership.

Committee on excursion reported, and the committee was discharged.

Voted, that the Secretary draw an order on the Treasurer for $5.00, to pay the deficit in expenses of the excursion in June.

Voted, that the Association make an excursion to the volcanic ash bed in the Berlin woods, and such expense be incurred as will be necessary.

Voted, that the Secretary draw his order on the Treasurer for $8.25, to pay bill for printing.

Voted, that Mrs. Kendrick make the necessary arrangements for the Berlin excursion.

Messrs. Davis, Platt and Ellsbree were appointed committee on programme for the October meeting.

Adjourned to meet Monday evening, October 10, 1887.

OCTOBER 10, 1887.

Regular meeting of the Association October 10, 1887.

Mrs. K. A. Hathaway was elected a member of the Association.

Bill of H. Wales Lines & Co., of $39.50, was ordered paid.

The President appointed Mr. A. B. Mather and Mrs. Kendrick to wait on Messrs. H. Wales Lines & Co. in regard to the bill.

Committee on lectures reported, and committee was discharged.

Letter communicating the death of Professor Spencer L. Baird was received and ordered to be placed on file.

Adjourned to meet Monday evening, November 14, 1887.

NOVEMBER 14, 1887.

Regular meeting of the Association Monday evening, November 14, 1887.

Miss Lillie Chapman was proposed for membership.

Voted, that one hundred postal cards be sent out for the December meeting.

Adjourned to meet Monday evening, December 12, 1887.

DECEMBER 12, 1887.

Regular meeting of the Association Monday evening, December 12. 1887.

Miss Lillie Chapman was elected a member of the Association.

Messrs. Cooper and Watkins and Mrs. French were appointed committee on programme for the February meeting.

Committee appointed to wait on H. Wales Lines & Co., reported and the committee was discharged.

Adjourned to meet Monday evening, January 9, 1888.

1888.

Regular meeting of the Association Monday evening, January 9, 1888.

Mr. A. H. Andrews was proposed for membership.

The following were elected officers of the Association for 1888 :

President—DR. J. H. CHAPIN.

Vice-President—REV. J. T. PETTEE.

Secretary—DR. C. H. S. DAVIS.

Treasurer—REV. S. HALSTED WATKINS.

Curator and Librarian—ROBERT BOWMAN.

Mrs. F. Platt and Mrs. Kendrick were appointed a committee to revise the list of Directors.

Mrs. Ella B. Kendrick was appointed collector for 1888.

Article III of the By-Laws was amended so that two Auditors shall be appointed each year by ballot.

Messrs. H. S. Pratt and Geo. L. Cooper were elected Auditors for 1888.

Voted, that Mr. Pratt have the use of the cabinets for the use of the High School.

Voted, that the Auditors audit the Treasurer's accounts every six months.

Voted, that the Curator procure one or more cabinets for the Association.

Voted, that all persons who refuse to pay their annual dues be dropped from the Association.

Messrs. Davis, Pratt and Cooper were appointed a committee to revise the By-Laws of the Association.

Voted, that the committee on the purchase of books for the Association, have power to add to their number.

Adjourned to meet Monday evening, February 13, 1888.

FEBRUARY 13, 1888.

Regular meeting of the Association Monday evening, February 13, 1888.

Miss W. H. Hall was appointed Secretary, *pro tem.*

Mr. A. H. Andrews was elected a member of the Association.

Mr. C. V. Helmschmied was proposed for membership.

The report of the Treasurer was accepted and an itemized account called for.

Committee or the revision of the list of members reported progress, and the committee was continued.

Committee on the revision of By-Laws reported, and the committee was continued.

Adjourned to meet Monday evening, March 12, 1888.

MARCH 12, 1888.

Regular meeting of the Association Monday evening, March 12, 1888.

Carl Helmschmied was elected a member of the Association.

W. G. Riggs and J. F. Ives were proposed for membership.

Committee on revision of the By-Laws was continued to report at the April meeting.

Treasurer's report was read and approved.

Rev. Mr. Watkins, Mrs. Kendrick and Miss W. H. Hall were appointed a committee on programme for the June meeting.

Mrs. Kendrick and Miss Moses were appointed a committee to prepare resolutions on the death of Mrs. Pratt.

Adjourned to meet Monday evening, April 9, 1888.

APRIL, 9, 1888.

Regular meeting of the Association Monday evening, April 9, 1888.

W. G. Riggs and J. F. Ives were elected members of the Association.

Mr. and Mrs. W. C. Homan were proposed for membership.

Committee on revision of By-Laws reported, and 500 copies were ordered to be printed.

Voted, that the By-Laws go into effect at once.

Messrs. Cooper and Bowman and Mrs. Kendrick were appointed committee on excursion.

Adjourned to meet Monday evening, May 14, 1888.

MAY 14, 1888.

Regular meeting of the Association Monday evening, May 14, 1888.

Messrs. R. E. Chapman and W. J. Prouty were proposed for membership.

Mr. and Mrs. W. C. Homan were elected members of the Association.

Voted, that the bill for cabinet be paid.

Voted, that the bill for rent of hall for Mr. Draft's lecture be paid.

Voted, that Mr. Quested's per centage for collecting be paid.

Adjourned to meet Monday evening, June 11, 1888.

JUNE 11, 1888.

Regular meeting of the Association Monday evening, June 11, 1888.

R. E. Chapman and W. J. Prouty were elected members of the Association.

Voted, that the bill of Peck Brothers of $1.90 be paid.

Mr. Charles E. Ellsbree was elected Treasurer of the Association.

Messrs. Wheelock and Stone and Miss Lindsley were appointed committee on programme for the October meeting.

Adjourned to meet Monday evening, September 10, 1888.

SEPTEMBER 10, 1888.

Regular meeting of the Association Monday evening, September 10, 1888.

Mr. T. S. Frost was proposed for membership.

Mr. Geo. L. Cooper was appointed Director of the department of Mechanics.

Voted, that the By-Laws of the Association be referred back to the committee for correction.

Adjourned to meet Monday evening, October 8, 1888.

OCTOBER 8, 1888.

Regular meeting of the Association Monday evening, October 8, 1888.

Mr. T. S. Frost was elected a member of the Association.

Voted, that the bill of E. M. Judd, to the amount of $23.45 for books, be paid.

Miss Alice H. Derby was elected Auditor in place of Mr. H. S. Pratt.

Mrs. Kendrick, Mrs. Platt and Mr. Chapman were appointed committee on programme for the December meeting.

Voted that a committee of three be appointed by the President to prepare the third volume of the Transactions of the Association. Dr. Davis, Mr. Pratt and Mrs. Kendrick were appointed publishing committee.

Adjourned to meet Monday evening, November 12, 1888.

NOVEMBR 12, 1888.

Regular meeting of the Association, Monday evening, November 12, 1888.

Report of the committee on By-Laws accepted.

Report of the committee on programme accepted and committee continued.

Adjourned to meet Monday evening, December 10, 1888.

DECEMBER 10, 1888.

Regular meeting of the Association Monday evening, December 10, 1888.

Messrs. Frost and Barker and Mrs. Ellsbree were appointed committee on programme for the February meeting.

The Treasurer was instructed to pay the bill of The Eaton & Peck Co., of $5.00, and H. N. Brooks & Co., $2.00.

Adjourned to meet Monday evening, January 14, 1889.

CHAS. H. S. DAVIS,

Secretary.

The Ash Bed at Meriden and its Structural Relations.

By WILLIAM MORRIS DAVIS.

During a geological excursion in the Easter recess of 1887, with several students from Harvard College, we were walking down the Berlin road towards Meriden one afternoon, when our attention was taken by the peculiar appearance of a bold outcrop that made part of the trap ridge anterior to Lamentation Mountain. We had already examined the ridge a mile farther north, where it consisted of trap of the ordinary kind, and also at an intermediate point, where, as was long ago described by Percival in his State Report, the dense trap is for a distance of several hundred feet replaced by a conglomerate of more or less water-worn trap fragments, in which are intercalated thin beds of sandstone giving a clear idea of the dip of the whole deposit But the part of the bluff now reached looked like neither of the other outcrops, and we made our way up to it through the underbrush. A closer inspection showed it to consist of a gray, greenish mass of vague texture, much weathered on the surface, through which were scattered great rounded blocks of dense trap, very fine grained at the surface, and of all sizes from six inches up to several feet in diameter. It could hardly be interpreted otherwise than as an ash bed, into which blocks of lava had fallen during the shower of ashes that had formed it. One of the first blocks to fall indented the sandy mud on which the ash bed was accumulated.

Similar ash beds are described by Scrope in his classic volume on the Volcanoes of Central France, and are often mentioned in accounts of recent volcanoes. Lava blocks of large size are known to be thrown to considerable distances; Humboldt describes them and recent reports from Italy give the same story.

The ash bed is interesting enough in itself, but its value is still greater in the evidence that it gives in regard to the general history of the region. I gladly avail myself of the invitation of the Meriden Scientific Association to write a few pages about it, from this point of view. It is like a monument whose value is less in recording the facts of its construction than in recalling the events that it commemorates.

The ash bed is a good teacher on two matters of importance in the geological history of Meriden. It tells much in regard to the conditions of accumulation of the series of deposits that form the rock foundation of the country, and it tells something also about the deformation that the mass has suffered since its materials were brought together.

All observers are familiar with the reddish sandstones, shales and conglomerates that make the bed-rock of the country, and with the sheets of trap that form the ridges. In this part of the valley the trap does not occur in the form of dikes, but as sheets, in just the same interbedded position that beds or sheets of conglomerate might have. The question then arises, how did these sheets of trap attain their position between the beds of sedimentary deposits, and to this there are two answers. The trap sheets may have been shoved or intruded between the sedimentary rocks after the latter had been made, or they may have been poured out as lava flows during the making of the sedimentary series. In the first case they are called intrusive sheets, and are known to be such by the signs of baking that they produce on the over-lying beds, and by the dense texture at their upper surface, where they were under great pressure of the rocks above, as well as by the occasional appearance of small offshoots which cut the adjacent strata. In the second case, they are called extrusive, and are known to be such by the appearance of various features characteristic of lava flows, such as vesicular or scoriaceous upper surface of loose texture, by their association with ash beds, and by the occurrence of their fragments in sandy or muddy deposits lying near their margin or laid upon their surface. The two cases should be clearly conceived in order that the real value of the Meriden ash bed, as a point of evidence in the history of the valley, may be appreciated, and that it should not be regarded merely as a locality for collecting specimens. With these points in mind, the ash bed takes on a new meaning. We may see it, not simply as an outcrop in the face of a bluff, but as the present edge of a sheet or bed, of which much still remains under ground, not yet reached by the destructive forces of the weather, but of which another large part has already been thus consumed. The destructive process is still going on, and probably about as fast as it has worked on the average in the past. Now following along the edge of this sheet to the north, we find successively the ash bed, outcrops of dense trap, then bluffs of trap conglomerate, and further on, more dense trap, from which it may be safely argued that a bed or sheet which presents these varied appearances can only be of extrusive origin. Further evidence to the same conclusion will be found by examining a small outcrop on the back or eastern slope of the anterior ridge, about east of the ash bed, where the trap sheet is just covered by the over-lying sandstones, which near the trap contain fragments of it, showing unmistakably that the trap was there as a foundation on which these over-lying beds were deposited.

Similar characteristics may be found in various other trap ridges of the neighborhood, and constitute most interesting features for discovery, observation and reflection. I may instance one locality where the contact of the over-lying sandstone on the back of the upper surface of the thick trap sheet that forms Lamentation Mountain is particularly well shown; this is at Spruce Brook, where it runs on the eastern slope of Lamentation northward towards the road that passes by the north end of the mountain; it flows from the trap to the sandstone and exposes an excellent natural section that deserves as much attention as the ash bed, even though it is not quite so striking in its appearance. It shows the over-lying sandstone to contain many vesicular fragments derived from the trap, and in the vesicles of these fragments, a lens reveals the finer sediments lying just as fine sandy deposits lie in the scoriaceous surface of submarine lava-flows at the present time. The conclusion that these lava sheets are great surface flows, extruded at the time when the Triassic sandstones were accumulating, is borne out not only by the direct evidence of the kind here adduced, but also by the contrast thus afforded with the features of the West Rock ridge, which runs northward from New Haven. Roaring Brook, not far west of Cheshire, on the eastern slope of Gaylord's Mountain, has exposed a contact of the trap sheet of this mountain with the sandstone that lies on its back ; and the two together exhibit all the characteristics that might be expected of an intrusive sheet. The trap is dense and fine-grained, and not in the least vesicular; its upper surface is relatively smooth and the sandstone lying close upon it is indurated; at one point, a narrow, fine-grained branch or off shoot of the main sheet may be seen penetrating the over-lying beds for several feet; but these over-lying beds do not contain trap fragments. This association of features in the two cases makes it clear enough that intrusive sheets can be distinguished from extrusive sheets by their physical character- istics, and with a great degree of certainty. In the Meriden dis- tricts, all the sheets can be shown to be of extrusive origin.

When we picture the conditions of the estuary in which the Tri- assic deposits were accumulated, we must therefore imagine the or- dinary processes of erosion from the adjacent lands and deposition in the waters to have been interrupted occasionally by eruptive action on a large scale. At one time, a sheet of lava whose edge is now visible as the anterior ridge of Lamentation, was poured out over the muddy bottom of the estuary, and accompanying this eruption there was explosive action that gave rise to showers of ashes and bombardment of lava blocks. About the same sort of eruption, but of later date, is known in Massachusetts, on the back of Mt. Holyoke, where it has been carefully studied by Professor Emerson of Amherst. There was then a time of quiet deposition again for a period long enough to form the strata that lie between the ante- rior ridge and the main trap sheet of Lamentation Mountain; and after this, came the great out-pouring of lava to a thickness of sev- eral hundred feet and over an area that can be shown to measure several hundred square miles. No explosive action producing ashes and bombs has been detected in connection with this eruption. Another period of quiet deposition followed, succeeded by a smaller eruption, whose edge is seen posterior to Lamentation in the trap ridge which runs north from Highland Lake. There were probably other eruptions at yet later dates. The intrusion of the West Rock sheet cannot be dated with respect to the eruptions.

The opportunity for the deposition of the Triassic strata came when the old land surface on which they rest was depressed below water level, so that it received the waste from the adjacent unsub- merged areas. The deposition ceased when a later disturbance ele- vated the submerged area above water level and exposed its accu- mulations to destructive action.

The character of this disturbing force was very peculiar. It re- sulted first, in breaking up the whole series of deposits, aqueous and igneous together, into relatively narrow blocks from a quarter of a mile to one or two miles or more wide and ten or twenty miles long; and then in dislocating them by tilting them over so that the beds, which were at first horizontal, now incline or dip to the eastward at an angle of about fifteen degrees. The country would have a strange topography if this constructional form had never been changed by erosive forces. A pretty close parallel to it may now be found in southern Oregon, where the dislocations are of a type very similar to those of the Connecticut valley, but where the date of dislocation is so recent that there has not yet been time to produce much change in the constructional form. Meriden readers should not fail to study the excellent description of this remarkable western region, as presented by Russell in the Third Annual Report of the United States Geological Survey; it has a distinct bearing on the early history of their own country.

But the date of the dislocation in the Connecticut valley is remote. Time enough has since then passed to allow the erosive forces to smooth down the whole irregular surface of the broken blocks and reduce their original mountainous form to what has been called a plain of base-level denudation; that is, to a surface of moderate relief, worn down to the level of the standing waters into which the streams discharged. This was accomplished long ago, when the land mass stood lower than it does at present, and the old base-level plain is now to be seen only in such remnants as the elevated plateau of the crystalline rocks on either side of the valley and in the crest-lines of the higher trap ridges. For in consequence of the elevation that the old plain suffered at some time when its base-levelling was well advanced, it has since been again attacked by erosive forces, and is now on its way to be again reduced to a lower base-level plain; but time enough has not yet elapsed to allow the completion of this second cycle of work. The work is pretty well advanced on the softer rocks, such as the sandstones, but the crest-lines of the thicker trap sheets are as yet not much reduced below the level of the earlier base-level, and the great mass of hard crystalline rocks on either side of the Triassic belt is very slightly consumed except along the valley lines, where the fastest cutting is always done at first.

Now in order to understand the present topography, the reader must make a purely geometrical effort; he must picture the series of beds of sandstones and traps, broken into long, narrow blocks and tilted over, and reduced by erosion to a surface of very moderate relief, a base-level plain; and on this plain he must perceive that according to the amount of dislocation of the blocks and the amount of inclination given to their beds, any given member of the series, as the heavy sheet of trap already described in Lamentation Mountain, will appear in various outcrop lines. If there be any difficulty in understanding this, let the structure be imitated by laying a num ber of books of different breadths but of the same thickness beside each other, these representing the Triassic series in its original undisturbed position. The lengths of the books should be imagined of indefinite extension. Let pages 90 to 100 of every book represent the heavy trap sheet of the second eruption; then lift the left hand side of every book, corresponding to western side of the blocks, so that the book covers will slope fifteen degrees to the east. In order to give a closer imitation of the real occurrence, the books should lie with the dividing lines between them trending to the northeast, and should then be lifted at the western corners, so that the direction of their slope is eastward and not at right angles to the lines of dislocation by which they are separated. The uneven surface thus produced corresponds to the constructional form that the Triassic formation would have had when first disturbed, if no erosive forces had worked upon it during its disturbance. Now imagine the books worn down to a certain level, corresponding to the base-level of the first cycle of erosive work; and on this level surface, pages 90 to 100 of all the books would have several outcrops. It is further manifest that on such a base-levelled surface the sequence of pages can be found in proper order only by following a line that crosses the pages of a single book; in other words, the true sequence of Triassic deposits can be found only by crossing the country in a line between the faults that bound any given block; otherwise, in passing from one block to another, the line would traverse beds out of their normal order; and a single bed might be met as often as the line passed into a new block. On this geometrical principle depends the method that must be employed in unravelling the Triassic structure; the observer must keep between a pair of faults that enclose a block, if he would not confuse his section by encountering the members of the series out of their proper order.

It is simply impossible to understand the structure of the region and the problem that it presents without a clear comprehension of its geometrical relations, somewhat after the manner given here.

Now after the elevation of the old base-level plain, the erosive forces begin a new cycle of destructive work upon it; they wear down the sandstones rather quickly, but the trap sheets are more obdurate and withstand the weather more successfully. This is especially true of the heavy main sheet, which has not as yet lost much of the form that it had in the late stages of the history of the old plain, although it has gained relief by the lowering of the adjacent country. Wherever there was an outcrop of the main trap sheet on the old plain, there we now have a ridge developed; and inasmuch as the tilting and dislocation of the blocks into which the mass has been divided had produced repeated outcrops of the sheet on various lines, the present ridges will be arranged in accordance with these repetitions. Thus we can account for the numerous high trap ridges, such as Higby or Besick Mountain, east of Meriden, Chauncy Peak and Lamentation to the northeast, the Hanging Hills, and other equally prominent ridges further north and south, all of these being the edges of a single heavy lava flow, repeated in as many blocks, more or less dislocated. Here we find the evidence of the great original area of the main sheet. The course of the dislocations or faults by which the blocks are separated is about N. 60° E. in the Meriden region; for example, the fault that cuts off the south end of the quarry ridge north of the Fair Grounds in Meriden (the easternmost member of the Hanging Hills group) passes obliquely across the Beaver Pond valley and cuts off the northern end of Lamentation Mountain. At the quarry, one may pass in a few steps from the main sheet to a bed of conglomeratic sandstone that belongs perhaps two thousand feet lower down in the normal series, merely by crossing the branch railroad track. Following the line of the track northeastward, successive conglomerate ridges may be seen to end one after another as they run north to the invisible fault line; and the anterior trap ridge of Lamentation disappears in the same manner, a manner that would be mysterious enough if it were not so systematically connected with the prevailing structure of the region. Faults of the same kind may be traced through Cat Hole, Reservoir Notch and other passes that interrupt the range north of West Peak; also through the notch between Lamentation Mountain and Chauncy Peak, and between the latter and Higby Mountain.

In every block thus defined, the whole series of deposits must appear. The lower conglomerates and sandstones first, then the anterior trap sheet, the shales overlying it, the main trap sheet forming the dominating ridges, and behind these more shales and a posterior trap sheet, somewhat thinner than the anterior. The ash bed that is associated with the anterior sheet in the Lamentation block can be seen in the same relative position in the Chauncy Peak block, and again near the north end of the anterior ridge of the Higby Mountain block. It may yet be found elsewhere. It is in this structural relation that the ash bed has its highest value.

Enough has been said to point out the importance of study of the region from the physical point of view. It is not sufficient to examine one locality and another, regarding each simply as furnishing geological specimens; the relations of the several parts must be studied out, and when proper interpretation is given to them the results will well repay the work that they have cost. The direction in which further work can be profitably turned is first in searching out all the possible outcrops that reveal contacts of the traps with the overlying beds, in order to discover all the evidence in regard to the origin of the trap sheets, and next to trace with as much precision as possible the many faults that traverse the district, which thus far have been detected chiefly in the neighborhood of the trap ridges alone. The faults doubtless have an extension of many miles, and some of them may even traverse the valley obliquely from side to side.

One of the most effective aids to such explorations will be the preparation of a good map, in which I trust the Association will feel a strong interest. The Massachusetts survey is now completed, and its contoured map-sheets are in course of publication. The field work of Rhode Island is also nearly or quite done. It is most desirable that Connecticut should follow as rapidly as possible in this excellent work, so that within a few years we may have a reasonably correct physical map of southern New England, on which the relief of the surface shall be duly portrayed. It will then be a much less difficult task to advance in the interpretation of the geological structure of the Meriden region.

In case some of the members of the Association should desire to read further on this subject, I take the liberty of referring them to two articles now in course of publication: one in the Seventh Annual Report of the U. S. Geological Survey, in which a mechanism that has been suggested to explain the monoclinal deformation of the region is discussed; an abstract of this was printed in the American Journal of Science two years ago; another describing the methods of work pursued by the Harvard Summer School of Geology during its sessions of 1887 and 1888 for the week of its stay at Meriden, in which maps and detailed itineraries of our excursions are given; this will appear in the Bulletin of the Museum of Comparative Zoölogy at Cambridge. A third paper is in preparation with the assistance of Mr. C. L. Whittle, who has accompanied me on many visits to the valley, in which the contacts of the trap and the overlying sandstones will be especially considered in relation to the intrusive and extrusive origin of the trap.

Cambridge, Mass., December, 1888.

The Nidus of Lunatia-heros.

By Frank J. Seidensticker.

Lunatia-heros.

Class—*Gasteropoda.*	Family—*Naticidæ.*
Sub-class—*Prosobranchiata.*	Genus—*Lunatia.*

Sand collars, or as the children of Cape Cod call them, Tom-cods' nests, we are told by the authorities, are found on sandy beaches along the entire New England coast, but much more sel-dom to the south than to the north of Cape Cod. The diversity of nomenclature given by the scientists of the past to these struc-tures seems to indicate an ignorance of their true nature. In Dr. A. A. Gould's Report on the Invertebrata of Massachusetts (State doc. 1870), he includes the following list of names given by differ-ent authorities:

Flustra arenosa. Ellis. Zooph. and also his Corallines.

Flustre areneuse. Lamouroux Polyp. flex.

Flustre Arenacee. Blainv. Dict. des Sc. Nat.

Eschara lutosa. Pallas. El. Zooph.

Eschara Millepora arenosa Anglica. Ray. Syn.

Alcyonium arenosum. Gmélin Syst. Nat.—Shaw Nat. Misc.

Discopora Cribrum. Lam. An. Sans Vert.

Nidus of Lunatia-heros.

The Nidus, when found damp on the beach, presents the appearance of a white rubber washer or diaphragm; somewhat resembling an inverted shallow bowl, open at the top and bottom and,
considering its real make-up, is fairly tough and flexible. Sizes
vary from two to six inches in largest diameter, but the shape is
curiously constant. When dried they still preserve their shape, but
are naturally quite brittle. According to general appearance they
are constructed entirely of sand held together by some glutinous
substance; but on holding a specimen against the light we find it
is closely stippled throughout with transparent points. The thick-
ness of the nest is between $\frac{1}{32}$ and $\frac{1}{16}$ of an inch, and on separating the inner from the outer surface we at once see the honey-
combed character of the interior.

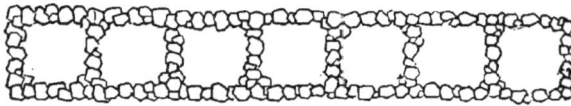

Diagram of Transverse Section of Nidus.

These interior cells are arranged in quincunx order, according
to Dr. Gould; and with a small lens and a strong transmitted
light this statement is readily verified. There is but one layer of
cells in the *thickness* of the Nidus. The interior of the cell ap-
proximates spherical in form, and each cell is perhaps a trifle less
than $\frac{1}{18}$ inch in diameter. Dr. Gould in his report states: "Each
of these cells contains a gelatinous egg having a yellow nucleus
which is the embryo shell." This hardly seems in accord with
the result of personal observation. Taking a transverse section of
the Nidus and examining the same with a power of 50 diameters,
one gets the decided impression that the greater portion of the
nest is made up not of sand but of the embryonic shells. The
surface and main portion of the walls between the cells consist of
layers of single grains of sand.

Embryonic Shell of Lunatia-heros.

The embryos or eggs are very pretty in shape, each measuring about $\frac{1}{100}$ inch in its largest diameter, each cell containing from one hundred to one hundred and twenty-five individual eggs. Quoting again from Dr. Gould's report : "The true nature of the Nidus seems to have been first suspected by Mr. Boys who gave a description and plate of it in the Linnæan Transactions, Vol. V. 230, pl. 10. In the fourteenth volume of the same work Mr. Hogg fully demonstrated its character by hatching from those found on the English coast the young of Natica glaucina."

Frequently attached to the inner surface of the Nidus are found colonies of little leather-like sacs of nondescript shape a little less than $\frac{1}{8}$ inch in length and about $\frac{1}{16}$ inch wide. These sacs are filled with shells almost exactly resembling those of the structural contents of the Nidus, but by micrometrical measurement appear to be a trifle smaller. Each sac contains over one hundred, and they are probably the eggs of some other mollusk which, cuckoo-like, has taken advantage of the industry of the Lunatia. In this connection it is interesting to remember that many of the mollusca, who at maturity are unprovided with a shell or who have only small suggestion of one imbedded in their mantel, have in their embryonic stage a perfect shell which they throw aside soon after the animal escapes from the sac. The mollusca, which were probably the earliest animal inhabitants of our globe, are often-times peculiarly constituted; both sexes existing in one individual. The *gasteropoda*, to which class the Lunatia belongs, are frequently hermaphrodite. (By the way, Mr. J. A. Ryder has established the remarkable fact that while in the American Oyster the sexes are separate the European Oysters are hermaphrodite.)

The Lunatia-heros, the builder of the Nidus, is one of our most common mollusks. The shell varies in size from $\frac{1}{2}$ inch upward. Dr. Gould states that he has seen one measuring five inches by three and three-fourths inches. This creature, which seems so exceptionally solicitous for the welfare of its own young, does not show like consideration for its neighbor. On the contrary, it is almost cannibal in its tastes. Dr. J. M. Crocker, of Cambridge, Mass., states that he has caught the Lunatia in the act of boring through the shell of a Quahaug in the most scientific manner, holding its prey so tenaciously that in picking up the Lunatia he lifted up Quahaug and all. It is supposed that the creature secretes an acid which softens the shell of its victim. But the main work of perforation is done by the *odontophore*, which is a cartilaginous strap armed with sharp transversely disposed teeth. This is worked backward and forward after the manner of a saw by the strong muscles to which it is attached. It is claimed that most of the round perforations so commonly found in shells cast on the beach are the work of the Lunatia.

Odontophore or Lingual Band of Lunatia-heros.

The lingual band of a Lunatia, mounted by Dr. Crocker, measures $\frac{1}{32}$ inch in width. There are in each row two excessively sharp curved teeth $\frac{1}{200}$ inch in length, besides three blunter median teeth.

The Lunatia is evidently the "Drill," who with the starfish and winkle have been doing injury among the New England oyster beds of sufficient magnitude to warrant the recent investigations of the representatives of the United States fish commission on board the U. S. Steamer "Fish-hawk."

Bodily removal in each case seems to be the only suggested cure of the evils.

The Trap Ridges of Meriden Again.

By J. H. CHAPIN, PH. D.

A protracted absence from the country, on a tour round the world, since the issue of the last volume of these Proceedings, has interfered with the exploration of the Trap dikes or ridges we had intended to make, and we have little therefore that is new to add to what has been said before.

Prof. W. M. Davis of Harvard University however, has vigorously prosecuted his explorations from time to time, and with his classes in the Summer School of Geology, has passed over portions of the region about Meriden repeatedly. His conclusions go far to strengthen and confirm the theory, that the whole area now occupied by the Trap ridges, was once covered by a continuous sheet of igneous rock, the result of one or more overflows; and that the separation of the ridges by intervening valleys or ravines, as at Cathole Pass, the Reservoir Notch, and the more considerable depression west of Lamentation Mountain was due to the rupture of the overlying sheet and faulting of the rocks. Prof. Davis speak for himself, however, in an article on another page.

The feature of more special interest at present is the evidence, of profound volcanic disturbance of the region named, in the bed of ashes and bombs of unmistakable origin, which stand out so prominently on the west slope of Mt. Lamentation—or speaking more precisely—in the Anterior ridge of Trap adjacent to Mt. Lamentation and near the road from Meriden through the Berlin Woods. The line of ash has been traced in a south-easterly direction and appears at different points along the range of ridges, leaving the observer quite in doubt as to the precise point of its origin. Whether there was one long rupture from which the ashes appeared, or a single crater of more pronounced volcanic character, or more than one volcanic vent along the flanks of these

Trappean hills, are points which as yet cannot be certainly determined.   Also whether there was a single period of disturbance in
which ashes and bombs appeared, or more than one, is a question
that awaits further investigation.   The disposition of the explorer,
and even of the casual visitor, now will be to look out for the
crater from which the material of the ash-beds came.   It may be
concealed beneath a considerable depth of rock or soil, in which
case the discovery will probably be purely accidental; still this
will be an item of interest until the point is located and the vent
if possible is found.

Some Notes of Africa.

In lieu of other matter, we venture to append a few notes made
in Northern Africa in May last, touching certain geological features of that country.

In the first place it may be stated in general terms, that the
northern coast of Africa, west of La Calle, is a series of headlands, with occasional intervening spaces sloping to the sea, and
with frequent indentations in the coast, that went far to adapt it to
the uses of the Turkish corsairs and the Algerine pirates, of which
the world has heard so much.   Further, the interior country,
especially in Algiers, is divided into three well defined belts, or
regions parallel to the sea and extending southward into the
desert.   They are known respectively as the TELL, the HIGH
PLATEAU, and the DESERT border.

The Tell is an undulating plain lying immediately back of the
headlands that mark the coast, and extending from fifty to eighty
miles in width, with a length east and west of perhaps seven hundred miles.   It is almost uniformly fertile and large portions of it
are under cultivation.   It is traversed in the eastern portion by a
spur of the Atlas Mountains and has also considerable areas of bog
or swamp.

The High Plateau consists of a series of benches or terraces,
lying nearly parallel to the Tell upon the one hand, and the
desert on the other, and sloping from a central ridge in both
directions.   Then comes the desert of Sahara—not an interminable

plain of shifting sands by any means, but a hot, dry, barren region with much variety of surface. The eastern portion is depressed often below the general level, and covered in part with a hard but brittle clay in place of sand; while the western portion consists in large part of rocky plateaus sometimes rising into mountains. The southern portion is represented as more uniform in character, but of that we cannot speak from observation.

The High Plateau, with gradient slopes both to north and south, and exposed therefore to the sea winds upon the one hand and the hot breath of the desert on the other, is of very doubtful character, both as to climate and productiveness. It is subject to great extremes of heat and cold—being three to four thousand feet in height in the central portion—and while it sometimes yields an abundant vegetation, at other times it fails almost entirely.

The Metija and Chelef plains are subdivisions of the Tell, having a north and south rather than east and west direction, and are most remarkable for a great depth of comparatively recent deposit, evidently due to the winds blowing from the direction of Sahara across depressions in the High Plateau. The depth of recent deposit in the Metija plain—the lower part of which however consists of pebbles and gravel and cannot therefore be of sub-ariel origin—is set down at five hundred to six hundred feet. This presents a problem which as yet has not been solved.

There is also a pecularity about many of the rivers of Northern Africa, especially of Algiers. Most of them rise well up in the regions of the High Plateau, and at the time of flood bear large quantities of black and muddy water far out to sea, but at other seasons have not force enough to reach the sea at all, but are swallowed up in the sinks and marshes along the lower portion of the Tell.

Two points in Algiers have a special interest for the scientific student. One lies near the borders of Tunis, the other well toward the boundary of Morocco.

On the way from an interior town to Bone upon the coast, we came upon the springs of Hammam Meskoutin, which in a small way reminded us of the Mammoth Hot Springs in the Yellowstone Park. The water is almost up to the boiling point, and strongly charged with lime, with a moderate percentage of iron. The principal spring at present rises in a shallow cove upon the hill side, and the deposit around is of almost snowy whiteness. The water descending in tiny cascades is arrested here and there, by little basins which immediately overflow and send their steaming contributions on their way, very much as at the greater springs already named.

Here and there about this spring are what were evidently geyser cones, two or three of the larger of which still retain their shape, but most of them have crumbled, and being covered with a thin soil and creeping vines, are overlooked by the average traveller. Some Roman baths, in one or two of which the water may still be used, indicate that the springs were made use of as long ago as the Roman occupation.

The other point named lies far to the westward.

The diligence will carry the traveler, who has the nerve and pluck to undertake the all-night journey, from Oran to a most interesting geological region near the village of Kleber. A gray and arid looking mountain, or rather high plateau, rises just beyond the town, the broad summit of which is an almost uninterrupted mass of marble breecia with a large percentage of iron, and a smaller portion apparently of manganese.

The indications are that originally the rock was white marble associated with iron ore and therefore discolored in parts; that by some tremendous movement of the earth, the rock was crushed or broken into small angular fragments, and then settling down nearly in the same position, the disintegrated mass was cemented by the infiltration of water bearing with lime, some coloring substance, as iron or manganese. While the fragments retain in good part their original color, the matrix formed by the flownig water is of a reddish, brown, or purplish hue, making a breecia scarcely excelled in beauty by any rock in the known world.

A portion of the marble in this mountain still retains its original character and position, while the breecia occupies another part. It seems probable therefore that only one side of the mountain was thus crushed and shaken up, though no line of division can be distinctly traced. It is much to be regretted, that there is not enterprise enough in Algiers to work this quarry and send out its rare treasures to the world. A few specimens may be found in the city of Algiers and a few in London, but for the most part the deposit lies there as it was formed.

J. H. C.

The Horizontal Moon.

By Rev. J. T. Pettee.

As known to every one the moon appears much larger on the horizon than on the meridian, though it is then four thousand miles farther from us; by instrumental measurement it is smaller, but to the naked eye larger. Why is this? Some have supposed that the denser atmosphere near the earth operates as a lens, and actually magnifies; but if this were so the amplification would appear in the telescope, as well as to the naked eye, and the instrumental measurements would be larger instead of smaller. What then is the reason?

This question used to be a good deal discussed by philosophers. Gassendus, Descartes and Hobbes all tried their hand at it; and one hundred and sixty years ago Bishop Berkely labored on it through fourteen pages of his Minute Philosopher. About 1660 the learned Dr. Wallis contributed an article to the British Philosophical Transactions, in which he explained the phenomena in this way:

We see the rising moon beyond all objects along the line of sight, and therefore think it farther off than it really is, and *this illusion* makes it appear larger. And this explanation (for the want of a better) has been accepted by astronomers. I doubt whether any astronomer was ever really satisfied with it. I am sure the ordinary observer is not. It implies a comparison which we never make when looking at the rising moon, and which the little child, or unlettered peasant, to whom it appears as much larger than the meridional moon as it does to us, is utterly unable to make. And then, as justly observed by Bishop Berkely, if we saw it from behind a wall which cut off all intervening objects, it would appear no larger than on the meridian; or, as we may observe, if looked at through a tube, or seen rising over the sea, or above a prairie.

But what *is* the cause of this optical delusion? We can only answer that by some optical law, not well understood, all objects seen against the sky horizontally, or slightly elevated, appear larger than when seen at greater elevations. A kite appears larger just as soon as it leaves the ground than it does when it is well up in the air; I mean with the same length of string. A captive balloon, which is allowed to rise a thousand feet, appears smaller at that elevation than on the ground, with you a thousand feet from it, or at such a distance as the hypothenuse of the angle made by the rope and the ground would require. A man on a spire, one hundred feet high, appears smaller than a man at the base of the spire, at the distance of the hypothenuse from you.

I remember, some fifteen years ago, when there were not so many houses in my part of the city as there are now; when there was nothing between my south verandah and Markham's Hill, one summer afternoon a hay-rigging drove on to the hill, and the horse, and wagon, and men appeared larger than they ought had they stood on my lawn. I was almost startled—it seemed like an apparition.

Since then I have frequently observed the same magnification of objects on West Peak. The other day, as I was looking that way with a small telescope, using a power of about forty, a dog came on to the peak, and he looked larger than the largest dog ought to forty rods from me; and men were correspondingly magnified, not by the telescope, but by the horizontal vision. I was using a power which brought the peak within forty rods, or the eighth of a mile, at which distance the largest dog in Meriden would not have appeared as large as the dog on the peak. Now I am not prepared to say just what it was that made the dog appear so large, but have no hesitancy in saying that it was the same thing that makes the moon appear so large when rising. I am fully satisfied that distant objects are magnified when viewed horizontally.

A Supplementary List of the Birds

Of Meriden, Conn.

As Taken and Identified Since the Publication

of the

List in Vol. II. of "Transactions of the Meriden Scienfific Association."

By Franklin Platt.

Empidonax flaviventris.
Yellow-bellied Flycatcher.
Found during the migratory seasons.

Melospiza georgiana.
Swamp Sparrow.

Dendroica Castanea.
Bay-breasted Warbler.
Only during the migrations.   Not common.

Seiurus motacilla.
Louisiana Water Thrush.
Rare.   Only during the migrations.

Geothlyphis philadelphia.
Mourning Warbler.
During the migrations.   Very rare.

Sylvania canadensis.
Canadian Warbler.
Spring and Fall migrant.

Dendroica virens, named in my original list as a spring and fall migrant, should be considered as a summer resident, as I have, since the publication of that list, found it breeding at Spruce Glen.

A List of the Butterflies

Of Meriden, Conn.

By Franklin Platt.

In arranging this list I have followed the classification and nomenclature used by Mr. C. J. Maynard in his " Butterflies of New England."* Except where due credit has been given I have named no species but what I have found in my own excursions, and can therefore vouch for the correctness of the list, if not for its completeness.

Family SATYRIDÆ.

Genus SATYRUS.

1. **Satyrus alope.** Very Common.

Genus NEONYMPHA.

2. **Neonympha canthus.** Not common.
3. **Neonympha eurytris.** Rather common.

Family DANAIDÆ.

Genus DANAIS.

4. **Danais Archippus.** Abundant.

Family NYMPHALIDÆ.

Genus NYMPHALIS.

5. **Limenitis disippus.** Very common.
6. **Limenitis ursula.** Common.

---

* " The Butterflies of New England, with Original Descriptions. Etc.," by C. J. Maynard, Boston, Mass.; Bradlee Whidden. 1886.

Genus **GRAPTA**.

7. **Grapta interrogationis.** Common.

8. **Grapta comma.** Common.

9. **Grapta progne.** Common.

10. **Grapta J. album.** Very rare. Mr. James Hayes has a specimen in his collection captured some years ago—the only one he has found in twenty years' collecting.

Genus **VANESSA**.

11. **Vanessa antiopa.** Common.

Genus **PYRAMEIS**.

12. **Pyrameis atalanta.** Rather common.

13. **Pyrameis huntera.** Common.

14. **Pyrameis cardui.** Common.

Genus **JUNONIA**.

15. **Junonia iavinia.** Very rare. Only one specimen, captured some years ago by Mr. F. D. Buess.

Genus **ARGYNNIS**.

16. **Argynnis idalia.** Rather common.

17. **Argynnis cybele.** Not very common.

18. **Argynnis aphrodite.** Not very common.

19. **Argynnis myrina.** Very common.

20. **Argynnis bellona.** Very common.

Genus **MELITEA**.

21. **Melitea tharos.** Very abundant.

22. **Melitea phaeton.** Rare. Mr. Thomas Hayes captured one last summer near the peat works, and Mr. James Hayes has three or four in his collection.

Family **LYCÆNIDÆ**.

Genus **THECLA**.

23. **Thecla calanus.** Not uncommon.

24. **Thecla humuli.** Not common.

25. **Thecla smilacis.** Rare. (I have never found but one specimen.)

26. **Thecla titus.** Not common.

Genus LYCÆNA.

27. **Lycæna pseudargiolus.** (Form neglecta.) Rare.

28. **Lycæna comyntas.** Very common.

Genus CHRYSOPHANES.

29. **Chrysophanus thoe.** Rare. (Only one specimen.)

30. **Chrysophanus americanus.** Very abundant.

Genus FENISECA.

31. **Feniseca tarquinius.** Rare.

Family PIERIDÆ.

Genus COLIAS.

32. **Colias philodice.** Very abundant.

Genus TERIAS.

33. **Terias lisa.** Rare.

Genus PIERIS.

34. **Pieris rapæ.** Very abundant.

Family PAPILIONIDÆ.

Genus PAPILIO.

35. **Papilio troilus.** Not common.

36. **Papilio cresphontes.** Very rare. Mr. James Hayes found a much worn specimen in 1887, and Mr. Buess informs me that he took one some ten years ago.

37. **Papilio turnus.** Rather common.

38. **Papilio asterias.** Very common.

Family HESPERIDÆ.

Genus EUDAMUS.

39. **Eudamus tityrus.** Rather common.

40. **Eudamus lycidas.** Rare. (Only one specimen.)
41. **Eudamus pylades.** Rather common.

Genus THANAOS.

42. **Thanaos persius.** Not very common.
43. **Thanaos lucilius.** Rather rare.
44. **Thanaos brizo.** Not very common.
45. **Thanaos juvenalis.** Rather common.

Genus PHOLISORA.

46. **Pholisora catullus.** Not uncommon.

Genus ANCYLOXYPHA.

47. **Ancyloxypha numitor.** Very common.

Genus PAMPHILA.

48. **Pamphila metea.** Rare.
49. **Pamphila zabulon.** Common.
50. **Pamphila sassacus.** Rather common.
51. **Pamphila leonardus.** Very rare. I have never captured but one specimen, nor have I noticed it in any other collection in this vicinity.
52. **Pamphila peckius.** Very common.
53. **Pamphila cernes.** Quite common.
54. **Pamphila metacomet.** Not uncommon.
55. **Pamphila verna.** Rare.

A List of the Forest Trees and Shrubs

To be found in Meriden, Conn.

By Chas. H. S. Davis, M. D.

I have endeavored in this article to give an account of all of the forest trees and shrubs that are to be found in Meriden.

There are persons now living who can remember when the greater part of Meriden was covered by a forest. The original forest has long ago disappeared, and in many places has been replaced by a second and sometimes by a third and fourth growth of trees, and the forest area of Meriden now consists to a great degree of coppice growth, which is cut in rotations of about thirty years, for fire-wood mainly.

There are in Connecticut 650,000 acres of woodland, or 21 per cent. Hitherto our forests have been destroyed for the sake of immediate pecuniary gain or convenience, with no regard to the future supply of a material so valuable and neccessary for almost all pursuits.

During the last few years great attention has been paid to the maintenance of timber supply by cultivation, and to the raising of forest trees from seed, their care and treatment in the nursery, and their permanent planting. In 1886 the legislature passed the following act to encourage the planting of forest trees:

SECTION 1. The governor shall annually in the spring designate by official proclamation an arbor day to be observed in the schools and for economic tree planting.

SEC. 2. Chapter forty-nine of the public acts of eighteen hundred and seventy-seven is hereby amended to read as follows: Whenever any person shall plant land in this State not heretofore woodland, the actual value of which at the time of planting does not exceed twenty-five dollars per acre, to timber trees of any of the following kinds, to wit, Chestnut, Hickory, Ash, White Oak, Sugar Maple, European Larch, White Pine, Black Walnut, Tulip, or Spruce, not less in number than twelve hundred to each acre, and such plantations of trees shall have grown to an average height of six feet, the owner of such plantations may appear before the board of relief of the town in which such plantation is located, and on proving a compliance with the conditions herein, such plantations of trees shall be exempt from taxation of any kind for a period of twenty years thereafter. (Approved March 31, 1886, Chapter XC.)

For some years I have made a study of our forest trees, and have collected cabinet specimens of nearly every tree mentioned in this article, showing the bark, sap, and heartwood of each species. When complete, it will find its proper place in the cabinet of our Association.

The full value has been obtained by a determination of the specific gravity and the ash of the dry wood.

I wish to acknowledge my indebtedness to Messrs. John P. Hall, Julius H. Yale, Nathan S. Baldwin. George and Charles I. Foster, and others who have assisted me in identifying the different trees and procuring specimens for my collection.

I. CONIFERÆ. THE PINE FAMILY.

This Order comprises some of the most magnificent trees known, and valuable for their timber as well as for their products, which include the turpentines, resins, pitch, tar, etc. The woody fiber of the plants of this Order, under a high magnifying power, exhibit peculiar circular disks or markings.

SECTION 1. THE PINE AND FIR TRIBE.

1. PINUS STROBUS. (LINN.)

White Pine. New England Pine.

This is one of our most valuable trees, growing from one hundred to a hundred and fifty feet high, with stem sometimes four feet in diameter. Wood light, soft, not strong, very close, straight-grained, compact, easily worked and susceptible of a beautiful polish. Comparatively free from turpentine. More largely manufactured into lumber, shingles, laths, etc., than that of any other tree. Grows rapidly on light, poor, sandy soils. Specific gravity, 0.3854; ash, 0.19.

2. PINUS RIGIDA. (MILLER.)

Pitch Pine.

A medium size tree, forty to seventy feet high, with stem one to three feet in diameter. Wood light. soft, not strong, brittle, coarse-grained, compact. It is full of resin, and generally so well studded with knots as to be of little value except for fuel. When self-planted on the poorest of sandy land it grows at the rate of an inch in diameter in three or four years in the first twenty-five years. Largely used for charcoal. Specific gravity, 0.5151 ; ash, 0.23.

3. PINUS RESINOSA. (AITON.)

Red Pine. Norway Pine.

This tree is not very common in Meriden ; it is often confounded with Norway Spruce. In Europe the name is given to quite another tree. It grows here as rapidly as Pitch Pine, and sometimes to a greater height, often as high as eighty feet. Wood light, not strong, hard, rather coarse grained, compact and quite durable. Very resinous. Used for all purposes of construction, flooring, piles, etc. Specific gravity, 0.4854 ; ash, 0.27.

4. ABIES CANADENSIS. (MICHAUX.)

Hemlock. Hemlock Spruce.

This was once a very common tree in Meriden, but is becoming rapidly thinned out. At first it is of slow growth and very delicate, requiring shelter, but when once started it grows with great rapidity, often reaching a height of nearly a hundred feet, with stem three to six feet in diameter. Wood light colored, very coarse-grained, soft, brittle, difficult to work. It is extensively employed for roof boards and sheathing, as it holds a nail well. The bark is rich in tannin, and is in great demand for tanning leather. From Hemlock is obtained *Canada Pitch* or *Gum Hemlock,* and is used very much in plasters. The Oil of Hemlock is used in liniment. A decoction of the bark is often used as an astringent. Specific gravity, 0.4239 ; ash, 0.46.

5. ABIES NIGRA. (POIRET.)

Black Spruce. Double Spruce.

Sometimes called *Red Spruce.* This is cultivated as an ornamental shade tree, although it often grows to the height of seventy-five feet. Wood light, soft, not strong, straight-grained, compact, satiny, sometimes reddish, although often nearly white. Used for piles, posts, railway ties, etc. Essence of Spruce, prepared by boiling the young branches, is used in the manufacture of spruce beer. Specific gravity, 0.4584; ash, 0.27.

6. ABIES BALSAMEA. (MARSHALL.)

Balsam Fir. Balm of Gilead Fir.

A quick-growing but short-lived tree, usually growing thirty to forty feet high, but sometimes sixty to seventy. Wood very light, soft, not strong, coarse-grained, compact, not durable, and but of little value. "Canada Balsam," or Balm of Fir, an aromatic liquid, also resin obtained from the tree by puncturing the vesicles found under the bark of the stem and branches, is used medicinally and in the arts. Specific gravity, 0.3819; ash, 0.45.

7. LARIX AMERICANA. (MICHAUX.)

American Larch. Black Larch. Tamarack. Hackmatack.

This is inferior to the European Larch (*L. Europea*). It is a slender tree, from twenty to fifty feet in height, although sometimes reaching a height of a hundred feet, with stem two feet in diameter. Wood heavy, hard, very strong, rather coarse-grained, compact and durable. Used in posts and fencing. Much attention is given in Europe to its cultivation, and it might be advantageously planted here on unproductive land. The inner bark is recommended in the treatment of chronic affections of the pulmonary and urinary passages. Specific gravity, 0.6236; ash, 0.33.

SECTION II. **THE CYPRESS TRIBE.**

8. THUJA OCCIDENTALIS. (TOURNEFORT.)

White Cedar. Arbor-Vitæ.

Usually a small tree, growing to the height of thirty to fifty feet. Not common in Meriden. There are many varieties in cultivation, some exceedingly dwarf, others tall and quite slender. Wood very light, soft, not strong, brittle, rather coarse-grained, compact and durable near the soil. Used for posts, fencing, railway ties and shingles. The distilled oil and a tincture of the leaves have been found useful in the treatment of pulmonary and uterine complaints. Specific gravity, 0.3164; ash, 0.37.

9. CUPRESSUS THYOIDES. (Linn.)

White Cedar.

This valuable tree is restricted to swamps, and grows from forty to eighty feet in height, and stem from two to three feet in diameter. The wood is reddish, light, soft, fine-grained, and very durable. There is probably no other wood that will yield so much valuable timber to the acre. It is largely used in boat-building, for woodenware, cooperage, shingles, interior finish, telegraph and fence posts, etc.

10. JUNIPERUS VIRGINIANA. (Linn.)

Red Cedar. Savin.

Shrubby, or a small tree, growing from twenty to fifty feet high. The heart-wood of reddish color, light, soft, not strong, brittle, very close and straight-grained, compact, easily worked, and very durable. Largely used for interior finish, cabinet making, posts, sills, and almost exclusively, for lead pencils. A decoction of the leaves is occasionly used as a substitute for savin cerata, and an infusion of the berries as a diuretic. Specific gravity, 0.4926; ash, 0.13.

11. JUNIPERUS COMMUNIS. (Linn.)

Juniper.

A low, straggling shrub or small tree, seldom more than ten or twelve feet high, although the most common in Meriden is the prostrate form. There is an immense number of varieties of this species in cultivation. The oil extracted from the berries is used as a diuretic.

Section III. THE YEWS.

12. TAXUS CANADENSIS. (Gray.)

American Yew. Ground Hemlock.

A prostrate evergreen shrub, the stem trailing on the ground, or just beneath the surface, to a distance of six or eight feet, sometimes from two to four feet high. The older botanists considered this a distinct species from the English Yew (*T. baccata*), but Dr. Grey and others considered it only a well defined variety.

II. CUPULIFERÆ. The Oak Family.

13. QUERCUS ALBA. (Linn.)

White Oak.

One of our finest and most valuable forest trees. Grows in Meriden from sixty to eighty feet high, with stem from two to five feet in diameter. Wood strong, very heavy, hard, tough, close grained, liable to check unless carefully seasoned, durable in contact with the soil. The wood is always in great demand for a variety of purposes, especially for agricultural implements, carriages, interior finish, cabinet making, etc. A decoction of the astringent inner bark is employed medicinally where astringents are indicated. Specific gravity, o.7470 ; ash, o.41.

14. QUERCUS MACROCARPA. (Michaux.)

Burr Oak. Mossy-cup Oak. Over-cup Oak.

A handsome tree with luxuriant foliage and remarkably large acorns. Grows from fifty to eighty feet high, with stem four feet or over in diameter. Wood heavy, strong, hard, tough, close grained, more durable in contact with the soil than other Oaks. It is of little value except for fuel. Specific gravity, o.7453 ; ash, o.71.

15. QUERCUS OBTUSILOBA. (Michaux.) VAR.
Q. STELLATA. (Wagenheim.)

Post Oak. Rough Oak. Barren White Oak. Iron Oak.

A medium sized tree, forty to fifty feet high. Wood heavy, hard, close-grained, compact, checking badly in drying. Resembles White Oak. Used for fencing, cooperage and fuel. Specific gravity, o.8367 ; ash, o.79.

16. QUERCUS BICOLOR. (Willdenou.)

Swamp White Oak.

A large tree, sixty to eighty feet high, and stem five to eight feet in diameter. Wood closely resembling the White Oak, heavy, hard, strong, tough, close-grained, inclined to check in seasoning. Specific gravity, o.7662 ; ash, o.58.

17. QUERCUS PRINUS. (LINN.) VAR. Q. MONTANA. (WILLDENOU.)

Rock Chestnut Oak.   Swamp Chesnut Oak.   Chestnut White Oak.

A medium to large tree, with reddish, coarse-grained wood, much inferior to White Oak.   Grows from fifty to ninety feet high, two to four feet in diameter.   Wood heavy, hard, strong, rather tough, close-grained, inclined to check in drying.   The bark is rich in tannin.   Spacific gravity, 0.7499; ash, 0.77.

18.   QUERCUS TINCORIA.   (BARTRAM.)

Black Oak.   Yellow-Bark Oak.   Quercitron Oak.   Yellow Oak.

A large tree, sixty to eighty feet or more in hight, and two to four feet in diameter, with a thickish, deeply furrowed, dark-colored epidermis, and a spongy, yellow inner bark.   Wood heavy, hard, strong, not tough, coarse-grained, liable to check in drying. Layers of annual growth marked by several rows of very large open ducts.   The inner bark is an article of commerce, under the name of Quercitron: and is exported in large quantities to Europe, where it is used in dying yellow.   The wood is extensively employed by coopers and carriage makers.   The inner bark is used medicinally as an astringent in hemorrhages, etc.   Specific gravity, 0.7045 ; ash, 0.28.

19.   QUERCUS COCCINEA.   (WANGENHEIM.)

Scarlet Oak.

There is some doubt whether this is really distinct from *Q. tinctoria.*   The leaves turn bright red or scarlet in late autumn.   Gray bark, rough, but not deeply furrowed.   Wood heavy, hard, strong, coarse-grained, sometimes quite tough, but variable in texture and value.   Spicific gravity, 0.7405 : ash, 0.19.

20.   QUERCUS PALUSTRIS.   (DUROI.)

Pin Oak.   Swamp Spanish Oak.   Water Oak.

A very handsome medium size tree, from forty to sixty or seventy feet high, and one to two feet in diameter, with numerous, rather slender horizontal or drooping branches, which are frequently very knotty.   Wood heavy, hard, very strong, coarse-grained, inclined to check badly in drying. Used for shingles, clapboards, construction and in cooperage. Specific gravity, 0.6938; ash, 0.81.

21. QUERCUS RUBRA. (Linn.)

Red Oak. Black Oak.

A very large and common tree, thirty to ninety feet high, and two to four feet in diameter. Wood heavy, hard, strong, coarse-grained, inclined to check in drying. It is of inferior value, but is used for barrel staves, chairs, clapboards, etc. Specific gravity, 0.6540 ; ash, 0.26.

22. QUERCUS ILICIFOLIA. (Wangenheim.)

Bear Oak. Black Scrub Oak.

A low, dwarf shrub, three to eight feet high. Found in very poor soil. This is a dwarf species of the *Q. falcata*, or Spanish Oak. I do not know that the Spanish Oak can be found in Meriden.

23. FAGUS FERRUGINEA. (Aiton.) VAR. F. SYLVATICA. (Walter.)

American Beech. Beech Tree.

Grows from forty to eighty feet high, with a thin, even-surfaced, whitish bark. Wood very hard and firm, tough, close-grained, susceptible of a very fine polish, and is next to the Hickory in value for fuel. Largely used in the manufacture of chairs, shoe lasts, plane-stocks, handles, etc. Is of rapid growth, but generally considered short-lived. Specific gravity, 0.6883 ; ash, 0.51.

24. CASTANEA VULGARIS. (De Candolle.) C. VESCA. VAR. AMERICANA. (Michaux.)

Chestnut.

A variety of the European Chestnut. One of our most valuable forest trees, growing from sixty to ninety feet high, and from two to four feet in diameter. The tree is of rapid growth, being speedily reproduced by suckers from the stump. Wood light, soft, not strong, coarse-grained, liable to check and warp in drying, easily split. Is of a light yellowish or brown color, and is much used in the manufacture of furniture, fence rails, beams and inside finish. It makes very poor fuel, not worth half as much as hickory, as it burns slow, snaps disagreeably, and throws out little heat. An infusion or fluid extract of the dried leaves is employed in the treatment of whooping cough. Specific gravity, 0.4504 ; ash, 0.18.

25. CORYLUS AMERICANA. (MARSHALL.)

American Hazel.   Hazel Nut.   Wild Filbert.

A small shrub, four to six feet high   There is another native species, the Beaked Hazel, *C. rostrato*, which has the involucre prolonged into a bristly beak, extending an inch beyond the nut.

III.   CARPINACEÆ.   THE HORNBEAM FAMILY.

26. CARPINUS AMERICANA. (MICHAUX.) VAR.
C. CAROLINIANA. (WALTER.)

Hornbeam.   Blue Beech.   Water Beech.   Iron Wood.

Shrubs and trees, from twenty to forty feet high, often branched from the root, and growing in clusters.   A tree of slow growth and readily distinguished by its peculiarly rigid trunk.   Wood heavy, very strong and hard, close-grained, inclined to check in drying. Once extensively used for making brooms, as the wood is so tough that it can be split in narrow strips.   Specific gravity, 0.7286 ; ash, 0.83.

27. OSTRYA VIRGINICA. (WILLDENOU.)

Hop Hornbeam.   Iron Wood.   Lever Wood.

This is a handsome tree, from twenty to fifty feet high and from five to ten inches in diameter.   Wood heavy, very strong and hard, tough, close-grained, compact, susceptible of a beautiful polish.   Used for making beetles, mallets, mauls, tool handles, etc. Specific gravity, 0.8284 ; ash, 0.50.

IV.   JUGLANDACEÆ.   THE WALNUT FAMILY.

28. JUGLANS CINEREA. (LINN.)

Butternut.   White Walnut.

A large and rapid growing tree, from twenty to fifty feet high and two feet or more in diameter, readily raised from the nut, and can be safely transplanted at almost any age.   Wood light, soft, not strong, rather coarse-grained, compact, easily worked, satiny, susceptible of a beautiful polish. Largely used for interior finish, cabinet work, etc. The bark as well as the husks of the fruit is sometimes used as a dye. The inner bark, especially that of the root is employed medicinally as a mild cathartic. Specific gravity, o.4086; ash, o.51.

29. JUGLANS NIGRA. (Linn.)

Black Walnut.

This beautiful tree is becoming very scarce in Meriden. It grows from forty to eighty feet high, and from four to six feet in diameter, and is of rapid growth. When prevalent, is a pretty sure indication of a fertile soil. Wood of a dark, rich brown color, rather hard and firm, but susceptible of a high polish, rather close-grained, liable to check if not carefully seasoned, easily worked. Probably more extensively employed for cabinet work, gun stocks, interior finish, than any other native wood. Specific gravity, o.6115; ash, o.79.

30. CARYA ALBA. (Nuttall.)

Shell-bark Hickory. Shag-bark Hickory.

A large tree, often eighty feet high, and stem two to three feet in diameter. I think there are some varieties of this tree with nuts of a thicker shell, and the kernal of inferior quality, resembling the *C. suclata*, found in the west. A noble and valuable forest tree. Wood heavy, very hard and strong, tough, close-grained, flexible, and largely used in the manufacture of agricultural implements, carriages, axe handles, baskets, etc. Specific gravity, o.8372; ash, o.73.

31. CARYA TOMENTOSA. (Nutall.)

Mockernut. Black Hickory. Bull-nut. Big-bud Hickory. White-Heart Hickory. King-nut.

A very tall but slender tree, sixty to eighty feet or more in height. The bark with the fibers interlocked and not exfoliating. Is considered the best of the Hickories for fuel. Wood heavy, very hard, strong, tough, very close-grained, checking in drying, flexible. Is best cut in the month of August. Specific gravity, o.8216; ash, o.06.

32. CARYA PORCINA. (Nutall.) VAR. C. GLA-BRA. (Torrey.)

Pig Nut. Brown Hickory. Black Hickory. Switch-bud Hickory.

A large tree with smooth bark, forty to seventy feet high. Wood similar to *C. tomentosa*, and used for same purposes. Specific gravity, 0.8217; ash, 0.99.

33. CARYA AMARA. (Nuttall.)

Bitter-Nut. Swamp Hickory.

A small slender tree. Wood rather soft, white, but often quite tough, checking in drying. Kernal of nut intensely bitter. Used for hoops, ox-yokes, etc. Specific gravity, 0.7552; ash, 0.03.

V. BETULACEÆ. The Birch Family.

34. BETULA LENTA. (Linn.)

Cherry Birch. Black Birch. Sweet Birch. Mahogany Birch.

A large tree, thirty to sixty feet high, and one to two feet in diameter. Wood is colored reddish, heavy, very strong and hard, close-grained, compact, susceptible of a beautiful polish. Used in making cabinet ware, bedsteads, and for fuel. Birch beer is obtained by fermenting the saccharine sap of this and perhaps some other species of the genus. Specific gravity, 0.7617; ash, 0.26.

35. BETULA EXCELSA. (Pursh.) VAR. B. LUTEA. (Michaux.)

Yellow Birch. Gray Birch.

One of the largest and most valuable deciduous trees, growing to the height of seventy or eighty feet. Readily distinguished from *B. lenta* by its yellowish silvery or pearly bark. Wood, heavy, very stong and hard, very close-grained, compact, satiny, susceptible of a beautiful polish. Used for fuel, furniture, button moulds, pill and match boxes, and the hubs of wheels. Specific gravity, 0.6553; ash, 0.31.

36. BETULA NIGRA. (Linn.)

Red Birch. River Birch. Black Birch.

A small slender tree, though sometimes reaching from forty to sixty feet in height. The young trees and branches have a smoothish cinnamon-colored bark, the outer layers of old bark exfoliating in thin laminae or sheets. Wood, light, rather hard, strong, close-grained, and durable when not exposed to the weather. Valuable as fuel, and used in the manufacture of furniture, woodenware, etc, Specific gravity, 0.5792; ash, 0.35.

37. BETULA PAPYRACEA. (Aiton.)

Canoe Birch. White Birch. Paper Birch.

Does not grow to a large size in Meriden. It is closely allied to the White Birch, *B. populifolia*, but a larger tree. Branches slender and flexible, and the shining brown bark is dotted with white and readily separates into thin paper-like layers. Said to be the material of which the aborigines made their portable canoes, tents and baskets. Specific gravity, 0.5762; ash, 0.35.

38. BETULA ALBA. (Spach.) VAR. B. POPULI-FOLIA. (Marshall.)

White Birch. Old-field Birch. Gray Birch.

A small, short-lived tree of rapid growth, twenty or more feet high, with a chalky-white bark, and numerous slender branches. Wood, light and soft, close-grained, and liable to check in drying. Largely used in the manufacture of shoe-pegs, spools, wood pulp and charcoal. The bark and leaves are used medicinally. Oil is obtained from the inner bark by distillation. Specific gravity, 0.5760; ash, 0.29.

39. ALNUS SERRULATA. (Willdenou.)

Common Alder. Black Alder. Smooth Alder. Candle Alder.

Grows in swampy, meadows, and along streams, from three to ten or twelve feet high, and one-half to two inches in diameter. Wood, light, soft and close-grained. Decoction of the bark and leaves is used medicinally to purify the blood and for diarrhœa. Specific gravity, 0.4656; ash, 0.38.

40. **ALNUS INCANA.** (Willdenou.)

Speckled Alder. Hoary Alder. Black Alder.

This is distinguished from the common Alder by the polished appearance of its bark, and the whitened under surface of its leaves. Does not grow large enough to be useful, except for fuel and charcoal, and sometimes for gunpowder. Wood, light, soft and close-grained. Specific gravity, o.4607; ash, o.42.

VI. MYRICACEÆ. The Wax-Myrtle Family.

41. **MYRICA CERIFERA.** (Linn.)

Bayberry. Wax Myrtle.

Shrub three to eight feet high. Wood, light, soft, strong, brittle, very close-grained.   The leaves and bark of the root are used medicinally.   The wax which covers the fruit is called Bayberry talow, and was formerly used for candles, and in soap making. Specific gravity, o.5637 ; ash, o.51. *Myrica Gale*, (Sweet Gale ; Dutch Myrtle), is a bush found along the borders of ponds.

42. **COMPTONIA ASPLENIFOLIA.** (Aiton.)

Sweet Fern.

A very common shrub, and belongs to this Order.   An infusion of the leaves is uses in dysentery.

VII. PLATANACEÆ. The Plane Tree Family.

43. **PLATANUS OCCIDENTALIS.** (Linn.)

Buttonwood. Button-Ball. Sycamore. Water Beech.

One of the largest trees found in Meriden, from sixty to one hundred feet high and from two to five feet or more in diameter. Wood, brownish, cross-grained, cannot be split, and for this reason is used for meat blocks, and similar purposes. Decays quickly and is of very little value. Specific gravity, o.4880 ; ash, o.11.

VIII. SALICINÆ. THE WILLOW FAMILY.

44 POPULUS GRANDIDENTATA. (MICHAUX.)

Poplar. Large-toothed Aspen.

A large tree, sixty to eighty feet high, and from two to three feet in diameter, with rather smoothish gray bark. Wood, light, soft not strong, close-grained, and spongy, and of but little value. Largely mauufactured into wood pulp. Specific gravity, 0.4632; ash, 0.45.

45. POPULUS TREMULOIDES. (MICHAUX.)

Quaking Aspen. American Aspen.

A medium sized tree, fifty to seventy-five feet high, with stem twelve to twenty-four inches in diameter, with smooth bark. Wood white, soft, but of firm texture, somewhat resembling that of the White Birch, close grained and not durable. Makes excellent fuel, and is made into wood pulp. A bitter principle in the bark makes an excellent tonic in cases of debility. Specific gravity, 0.4032; ash, 0.55.

46. POPULUS BALSAMIFERA. (LINN.) VAR. P. CANDICANS. (GRAY.)

Balsam Poplar. Tacamahac. Balm of Gilead.

This is a rare tree, and is introduced and cultivated for ornament. Wood, very light, soft, not strong, close-grained, compact. The buds are used medicinally as a substitute for turpentine and other balms. Specific gravity, 0.3635; ash, 0.66.

47. POPULUS MONILIFERA. (AITON.) VAR. P. LÆVIGATA. (AITON.)

River Poplar. Cottonwood. Necklace Poplar.

A large tree, growing rapidly. Wood, very light, close-grained, liable to warp in drying, difficult to season; burns rapidly when seasoned, but gives out little heat. Used in the manufacture of paper pulp, light packing cases, fence boards and fuel.

48. SALIX TRISTIS and S. HUMULUS.

Sage Willow. Dwarf Gray Willow. Low Bush Willow.

Identified by Foster Brothers. Have never met with them.

49. SALIX DISCOLOR. (Muhlenberg.)

Bog Willow. Glaucous Willow.

A tall tree, or more often a tall, straggling herb. Wood, light, soft, close-grained ; color, brown, streaked with orange. Specific gravity, 0.4261 ; ash, 0.43.

50. SALIX ERIOCEPHALA. (Anderson.) S. PRIN-OIDES. (Pursh.)

Silky-headed Swamp Willow.

Identified by John P. Hall. These are varieties of the *S. discolor*.

51. SALIX CORDATA. (Mulhenberg.)

Heart-leaved Willow. Diamond Willow.

A small tree, but more often a straggling shrub. Wood, light, soft, close-grained. Very rarely in Meriden attains arborescent size or habit.

52. SALIX ALBA. (Linn.)

White Willow.

This is an introduced, but inferior, forest tree, and scarcely worth cultivating. The variety *S. vitellina*, Yellow Willow, or Golden Osier, has orange-yellow branches, and rather shorter and broader leaves, and is more extensively propagated than any other foreign Willow.

53. SALIX NIGRA. (Marshall.)

Black Willow.

A small tree from fifteen to thirty feet high, and not common in Meriden. It has a rough, black bark, and the wood is soft, weak and close-grained. The tonic and astringent bark is used medicinally, and contains, in common with that of all species of the genus, salicylic acid. Specific gravity, 0.4456 ; ash, 0.70.

54. SALIX LUCIDA. (Muhl.)

Shining Willow. Glossy Willow.

Rarely more than twenty feet high ; color, light brown ; wood, light, not strong, brittle and close grained. Specific gravity, 0.4547 ; ash, 0.79.

55. SALIX BABYLONIA. (Linn.)

Weeping Willow. Drooping Willow.

From thirty to fifty feet high. Is of but little or no economic value. There are others of this difficult genus, growing in low grounds, and mostly native species. There have been identified in Meriden, *S. rostrata*, Beaked Willow ; *S. fragilis*, Crack Willow ; *S. Russelliana*, Bedford Willow.

IX. ARTOCARPÆ. The Bread-Fruit Family.

56. MORUS RUBRA. (Linn.)

Red Mulberry.

From fifteen to twenty-five feet high, and nine to eighteen inches in diameter. Not common in Meriden. Wood, yellow, very heavy and durable, coarse-grained, satiny, susceptible of a good polish. Used in fencing, cooperage. snaths, etc. The leaves have been successfully used for feeding silk worms. Specific gravity, 0.5898 ; ash, 0.71.

The *M. nigra*, Black Mulberry, has been identified in Meriden by Foster Brothers and J. H. Yale, but I have not seen it. It is an introduced tree, as well as the *M. alba*, White Mulberry, of which a few specimens are found in Meriden. It should be cultivated more largely for feeding silk worms and the production of silk.

X. ULMACEÆ. The Elm Family.

57. ULMUS AMERICANA. (Linn.)

White Elm. American Elm. Water Elm. Weeping Elm.

A large and common tree, from sixty to eighty feet in height, and should be generally planted in every street in Meriden. Wood, brown, very tough in young trees, rather cross-grained, difficult to split. Used in the manufacture of hubs, saddle-trees, flooring, cooperage, etc. Specific gravity, 0.9506 ; ash, 0.80.

58. **ULMUS FULVA.** (Michaux.)

Slippery Elm.   Red Elm.   Moose Elm.

Grows from thirty to fifty feet high, and twelve to eighteen inches in diameter.   Wood, heavy, hard, strong, very close-grained, compact, splitting readily when green.   Used for wheel stock, fence posts, sills, railway ties, etc.   The inner bark is mucilaginous, nutritious, and extensively used in various medicinal preparations.   Specific gravity, 0.6956; ash, 0.83.

Some of the introduced Elms which have been identified in Meriden, but which are not common, are *U. campestre*, common European Elm; *U. montana*, Scotch Elm; *Celtis occidentalis*, Nettle Tree, and *C. cranifolia*, Hackberry.

XI.   LAURACEÆ.   The Laurel Family.

59. **SASSAFRAS OFFICINALE.** (Nees.)

Sassafrass.

Grows from fifteen to forty and fifty feet high, and six to twelve inches in diameter.   Wood, light, soft, not strong, brittle, cross-grained, slightly aromatic.   The root, and especially its bark, enters into commerce, affording a powerful aromatic stimulant; the oil is distilled from the root.   The pith of the young branches, infused with water, furnishes a mucilage used as a demulcent in febrile and inflammatory affections.

60.   **BENZOIN ODORIFERUM.**

Fence Bush.   Spice Bush.   Wild Allspice.

A shrub, four to ten feet high, and strongly aromatic.   Flowers in April.

XII.   OLEACEÆ.   The Olin Family.

61.   **LIGUSTRUM VULGARE.**   (Linn.)

Common Privet.

This is an introduced tree, and not common.   It grows from six to eight or ten feet high.   Many branches; branches opposite.   Berries, black.   Flowers in June, and fruit ripens in October.

62. SYRINGA VULGARIS. (LINN.)

Lilac.

This is a native of Persia, etc., and is one of our most common ornamental shrubs. Grows from ten to twenty feet high, in dense clumps. There are are several marked varieties.

63. FRAXINUS AMERICANA. VAR. F. ACU-MINATA. (LAMARCK.)

White Ash.

A large, handsome tree, with gray, furrowed bark on the main stem. Grows from seventy to eighty feet high, and from two to three feet in diameter. This is one of our most valuable timber trees, the wood being very tough and hard, ultimately brittle, coarse-grained. Is of rapid growth. Used in the manufacture of agricultural implements, carriages, oars, and for interior and cabinet work. Specific gravity, 0.6543; ash, 0.42.

64. FRAXINUS PUBESCENS. (LAMARCK.)

Red Ash.

A small, but rather slender, tree, in swamps and along streams. Grows from twenty-five to fifty feet high, and ten to eighteen inches in diameter. Is much less valuable than the *F. Americana.* Wood strong, brittle, coarse-grained, compact; color, rich brown. Specific gravity, 0.6251; ash, 0.26.

65. FRAXINUS SAMBUCIFOLIA. (LAMARCK.)

Black Ash. Hoop Ash. Ground Ash. Water Ash.

Mostly confined to swamps or wet soil. Grows from twenty-five to fifty feet high, and ten to eighteen inches in diameter. The wood is very tough, rather coarse-grained, separating easily into thin layers. Used in the manufacture of baskets, hoops, seating chairs, cabinet making, and interior finish. Specific gravity, 0.6318; ash, 0.72.

XIII. AQIFOLIACEÆ. THE HOLLY FAMILY.

66. ILEX OPACA. (AITON.)

American Holly.

A tree from fifteen to forty feet high. One of the most beautiful broad-leaved evergreen trees. Not common in Meriden. The wood is very compact, and of fine texture, tough, close-grained, and easily worked. Used in the manufacture of whip handles, screws, cabinet work, and interior finish. A bitter principle (Ilicin) is obtained from the fruit of this tree, and is used medicinally. Specific gravity, 0.5818; ash, 0.76.

67. ILEX VERTICILLATA. (Gray.)

Black Alder. Winter-berry.

A shrub, six to eight feet high. The bright scarlet berries in autumn and early winter make this shrub a very conspicuous object. Bark and berries used medicinally.

68. ILEX LÆVIGATA. (Gray.)

Single-berry Black Alder.

Found in swamps. The leaves are smooth beneath, the sterile flowers long-peduncled, and larger berries than the *I. Verticillata*.

69. ILEX GLABRA. (Gray.)

Ink-berry.

A delicate evergreen shrub, narrow leaves and black berries.

XIV. CAPRIFOLIACEÆ. The Honeysuckle Family.

70. LINNŒA BOREALIS. (Linn.)

Twin Flower.

A common creeping evergreen herb, with creeping, woody stem.

71. TRIOSTEUM PERFOLIATUM. (Linn.)

Fever-wort. Fever Root. Horse Gentian.

Grows from two to four feet high. A coarse looking plant, found in shady places. The root is used medicinally.

72. LONICERA HIRSUTA. (Aiton.)

Hairy Honeysuckle.

A hardy climbing plant, found in damp, rocky places. There has also been identified in Meriden the *L. parviflora*, small flowered Yellow Honeysuckle; *L. ciliata*, Fly Honeysuckle.

XV. VIBURNEÆ. THE ELDER FAMILY.

73. SAMBUCUS PUBENS. (MICHAUX.)

Panicled Elder.

This is the eastern variety of the *S. racemosa*, European Elder, which is common west of the Rocky Mountains.

74. SAMBUCUS CANADENSIS. (LINN.)

Common Elder. Elder bush. Black-Berried Elder.

This is the most common of the Elders found in Meriden. It grows from five to eight or ten feet high. An infusion of the flowers is an efficient diuretic, and the juice of the berries is made into wine. The bark is purgative and emetic.

75. VIBURNUM LENTAGO. LINN.)

Sweet Viburnum. Sheep-berry. Nanny-berry.

Grows from fifteen to twenty feet high. Wood, hard and yellowish, close-grained and strongly scented. Specific gravity, 0.7303; ash, 0.29.

76. VIBURNUM DENTATUM. (LINN.)

Arrow-wood.

A shrub, or small tree, very common in Meriden.

77. VIBURNUM ACERIFOLIUM. (LINN.)

Maple-leaved Arrow-wood.

A low shrub, common in rocky woods.

78. VIBURNUM OPULUS. (LINN.)

Cranberry Tree. Bush, or High, Cranberry.

Shrub. Grows from three to ten feet high, with spreading branches. This is the parent of the well known Guelder Rose, or Snowball, of gardens, in which the flowers are all sterile. The acid fruit is sometimes used as a substitute for cranberries, whence its common name. Flowers in June. Fruit in September.

79. VIBURNUM PRUNIFOLIUM. (Linn.)

Black Haw.   Stag Bush.

Not common in Meriden.   A large shrub, or small tree, ten to twenty feet high.   Wood, very hard, strong, brittle, close-grained. The bark is tonic and astringent, and in the form of decoctions and fluid extracts, is used in the treatment of uterine disorders. Specific gravity, 0.8332 ; ash, 0.52.

There are several other varieties of the *Viburnum* in Meriden, which I have not been able to identify.

XVI.   ERICACEÆ.   The Heath Family.

80. GAULTHERIA PROCUMBENS.   (Linn.)

Partridge Berry.   Tea-berry.   Checkerberry.   Wintergreen.   Box-berry.

A delicate evergreen plant, creeping on or near the surface of the ground.   The leaves are aromatic, and yield on distillation a very heavy volatile oil.

81. RHODODENDRON MAXIMUM.   (Linn.)

Great Laurel.   Rose Bay.

A tall tree or more often a tall, straggling shrub.   Not common in Meriden.   Wood heavy, hard, strong, close-grained.   Sometimes used as a substitute for boxwood in engraving.   Decoction of leaves used medicinally in the treatment of sciatica and rheumatism.   Specific gravity, 0.6303 ; ash, 0.36.

82. CLETHRA ALNIFOLIA.   (Linn.)

White Alder.   Sweet Pepper-bush.

Grows from three to ten feet high.   Flowers are exceedingly fragrant.

83. KALMIA LATIFOLIA.   (Linn.)

Mountain Laurel.   Calico Bush.   Spoon-wood.   Ivy.

Grows from three to ten feet high, with irregular, crooked, straggling branches.   Wood very hard, strong, close-grained, compact.   Leaves said to be poisonous to cattle.   The leaves, buds and fruit are used medicinally.

84. ARCTOSTAPHYLOS UVA-URSI. (Spreng.)

Bearberry.   Upland-Cranberry.   Uva Ursi.

A shrubby evergreen plant, trailing on the ground.   The leaves are used medicinally as a diuretic.   Flowers in May.

There are many other varieties in Meriden belonging to this family which have been identified: *Andromeda polifolia*, Water Andromeda, a low shrub, in boggy places; *Cassandra calyculata*, Dwarf Cassandra, a low evergreen, in swamps; *Zenobia racemosa*, Clustered Zenobia, a low shrub, much resembling whortleberry bushes; *Epigea repens*, Mayflower, a trailing evergreen; *Rhododendron viscosum*, Swamp Pink, Wild Honeysuckle, abundant in open woods.

XVII. VACCINIEÆ. The Whortleberry Family.

This is a Heath Sub-Family.   The following have been identified in Meriden:

85. GAYLUSSACIA RESINOSA.

Black Whortleberry.

86. GAYLUSSACIA FRONDOSA.

Dangleberry.

87. VACCINIUM CORYMBOSUM.

High-bush Huckleberry.

88. VACCINIUM VACILLANS.

Blue Huckleberry.

89. VACCINIUM PENNSYLVANICUM.

Low Blueberry.

90. OXYCOCCUS MACROCARPA.

Common Cranberry.

XVIII. CORNACEÆ. The Cornus Family.

91. CORNUS FLORIDA. (Linn.)

Flowering Dogwood.   Common Dogwood.   Dogwood.

A very beautiful tree, growing from fifteen to forty feet high, and three to eight feet in diameter.   The flowers appear in the spring, before the leaves. Wood very close-grained and firm, satiny, susceptible of a beautiful polish, checking badly in drying. Used for hubs of wheels, barrel hoops. The branches were formerly used as distaffs. The bark furnishes a valuable tonic, and is used in intermittent and malarial fevers. Specific gravity, 0.8153; ash, 0.57.

92.   NYSSA SYLVATICA. (Marshall.) VAR. N. MULTIFLORA. (Wang.)

Tupelo. Sour Gum. Pepperidge. Black Gum.

A handsome tree, growing as high as sixty feet, and from one to two feet in diameter. Wood strong, very tough, unwedgeable, fibres remarkably interlocked, so as to render it very difficult to split. Used for hubs of wheels, rollers in glass factories, ox yokes, etc. Specific gravity, 0.6353; ash, 0.52.

93. CORNUS ALTERNIFOLIA.

Alternate-leaved Cornel.

A shrub, sometimes twenty feet high, and five inches in diameter.

94. CORNUS CIRCINATA.

Round-leaved Cornel.

A shrub, seven to ten feet high.

95. CORNUS PANICULATA.

Panicled Cornel.

A slender plant.

96. CORNUS SERICEA.

Silky Cornel.

97. CORNUS CANADENSIS.

Dwarf Cornel.

XIX.   HAMAMELACEÆ. The Witch Hazel. Family.

98. HAMAMELIS VIRGINICA. (Linn.)

Witch Hazel.

Grows from six to twelve feet high. Flowers in October, the fruit perfecting in the summer following. Wood, heavy, hard, very close-grained, compact. The bark and leaves rich in tannin, and in the form of decoctions and fluid extracts is largely used medicinally. Specific gravity, 0.6856; ash, 0.37.

99. LIQUIDAMBER STYRACIFLUA. (LINN.)

Sweet Gum. Star-leaved Gum. Liquidamber. Red Gum. Bilsted. Alligator-tree.

Not common in Meriden. Resembles the Sugar Maple, but with a more conical head. Grows rapidly, and reaches a height of from fifty to seventy feet, and two or more feet in diameter. Wood fine-grained, inclined to shrink and warp badly in seasoning, susceptible of a beautiful polish; decays rapidly when exposed to the weather, and is of very little value as fuel. The balsamic exudation is obtained from the tree only in warm climates, and is used medicinally and also in the manufacture of chewing gums. Specific gravity, 0.5910; ash, 0.61.

XX. ROSACEÆ. THE ROSE FAMILY.

100. SPIRŒA OPULIFOLIA. (LINN.)

Nine Bark. Nine Bark Syringa.

Shrub, three to ten feet high, with spreading branches and loose lamellated bark, the numerous layers, suggesting the popular name. Flowers white, often tinged with purple.

101. SPIRŒA SALICIFOLIA. (LINN.)

Queen of the Meadows.

Abounds in wet places.

102. SPIRŒA TOMENTOSA. (LINN.)

Hardhack. Steeple bush.

A leafy shrub two to three feet high, in wet places. Stem and lower surface of the leaves covered with a rusty-colored wool. This plant is astringent, and is used medicinally in the treatment of diarrhœa and dysentery.

The *S. ulmifolia*, Meadow-Sweet, and *S. filipendula*, Drop-wort, and other well known ornamental plants belong to this genus.

103. RUBUS ODORATUS. (Linn.)

Rose-flowering Raspberry.

Stem perennial, three to five feet high, branching.   Fruit pleasantly flavored, but is rarely perfected under cultivation.   Flowers in June and July.   Fruit in July and August.

104. RUBUS STRIGOSUS. (Michaux.)

Wild Red Raspberry.

Stem from three to five feet high, light brown.   Flowers in May.   Fruit in July.   The fruit is collected for Raspberry Syrup.

105. RUBUS CANADENSIS. (Linn.)

Low Blackberry.   Dewberry.   Running Briar.

Stem from four to ten feet long, slender, trailing.   Flowers in May.   Fruit in July.

106. RUBUS VILLOSUS. (Aitkin.)

High Blackberry.   Common Brier.   Bramble.

Stem from three to six feet high, stout, rigid or angular, and somewhat furrowed.   The root is astringent and is used medicinally, as well as a wine made from the ripe fruit.   Flowers in May. Fruit in July and August.

107. ROSA CAROLINA. (Linn.)

Swamp Rose.

Grows four to five feet high, with numerous purple blossoms. Found in low, swampy grounds.   Flowers June and July.   Fruit in September.

108. ROSA LUCIDA. (Ehrh.)

Dwarf Wild Rose.   Early Wild Rose.

Differs from *R. Carolina* in its unequal bristly prickles and one to three-flowered peduncles.

109. ROSA RUBIGINOSA. (Linn.)

Sweet Briar.

Is well known for its fragrant glandular foliage.

The cultivated roses are varieties produced by long and careful culture from different species of this genus.

XXI. POMACEÆ. The Apple Family.

110. CRATÆGUS CRUS-GALLI. (Linn.)

Cock-spur Thorn. Newcastle Thorn.

Grows from ten to fifteen feet high; wide branches, and armed with short tapering thorns. Makes one of the most durable and effective hedges. Wood close-grained, hard, satiny, and susceptible of a fine polish. Specific gravity, 0.7164; ash, 056.

111. CRATÆGUS COCCINEA. (Linn.)

Scarlet-fruited Thorn. Scarlet Haw. Red Haw. White Thorn.

A small tree, seldom over twenty feet high. Leaves, bright green. Flowers, large, in large clusters. Wood hard, close-grained, compact. Specific gravity, 0.8618; ash, 0.38.

112. CRATÆGUS TOMENTOSA. (Linn.)

Pear-leaved Thorn. Black or Pear Hawthorn. Black Thorn. Pear Haw.

A common and hardy thorn. Grows from twenty to thirty feet high. Wood hard, not strong, close-grained. compact. Specific gravity, 0.7633; ash, 0.50.

There are a great many varieties of the Cratægus. The *C. oxycantha*, English Hawthorn, has been extensively introduced and cultivated in gardens. The *C. punctata* (Gray), Dotted-fruited Thorn, a variety of the *C. tomentosa*, is sometimes found in wet grounds.

113. PYRUS COMMUNIS. (Linn.) PYRUS MALUS. (Linn.)

Pear and Apple.

An extensive genus, containing about forty species, including the apple, pear, crab-apple, quince, etc., and innumerable varieties.

114. PYRUS AMERICANA. (De Candolle.)

American Mountain Ash.

Grows from ten to twenty feet high. A handsome, ornamental tree. Is very nearly related to *P. aucuparia*, European Mountain Ash. Wood light, soft. close-grained. Specific gravity, 0.5451 : ash. 0.83.

115. PYRUS ARBUTIFOLIA. (Linn.)

Choke-berry.

A slender, branching shrub.   Flowers white or tinged with purple.

116 AMELANCHIER CANADENSIS. (Torrey and Gray.)

June Berry.   Shad Bush.   Service Tree.   May Cherry.

A large shrub, or small tree.   Wood,. hard, very heavy, resembling that of the apple tree, susceptible of a good polish.   This species runs into many varieties, of which two have been identified in Meriden, the *A. botryapium*, June Berry, and *A. ovalis*, Swamp Sugar Pear.

XXII. AMYGDALEÆ. The Almond Family.

117. PRUNUS PENNSYLVANICA. (Linn.)

Wild Red Cherry.   Pigeon Cherry.   Pin Cherry.

Grows from twenty to thirty feet high.   Wood light, soft, close-grained, compact.   Fruit used medicinally.   Specific gravity, 0.5023 ; ash, 0.40.

118. PRUNUS PUMILA. (Linn.)

Dwarf or Sand Cherry.

Not common.   Identified by Foster Brothers.
A low, spreading or prostrate shrub, with many slender stems.

119. PRUNUS SEROTINA. (Ehrhart.)

Wild Black Cherry.   Rum Cherry.

Grows from forty to sixty feet high.   Wood light, strong, straight-grained, easily worked.   One of our most valuable woods for cabinet work and interior finish.   It takes a good polish, and is of a pale reddish tint, which deepens with age.   The bark is used medicinally in the form of infusions, syrups and fluid extracts.   The bitter fruit is used in the preparation of cherry brandy.   Specific gravity, 0.5822 ; ash, 0.15.

120. PRUNUS VIRGINIANA. (Linn.)

Choke Cherry.

A tall shrub.   Flowers in May.   Fruit in August.

XXIII. LEGUMINOSÆ. The Bean Family.

121. ROBINA PSEUDACACIA (Linn.)

Locust.   Black Locust.   Yellow Locust.   False Acacia.

Grows from thirty to sixty feet high, and from one to two feet in diameter.   Not a native here.   The timber is valuable for strength and durability.   Wood exceedingly hard and strong, close-grained. This tree is worthy of larger cultivation, as it grows rapidly on light and poor land.   The bark of the root is tonic and, in large doses, purgative and emetic.   Specific gravity, 0.7333; ash, 0.51.

122. GLEDITSCHIA TRIACANTHOS. (Linn.)

Honey Locust.   Black Locust.   Three-thorned Acacia. Sweet Locust.

This tree is introduced, and is often quite large and handsome, and is not liable to be attacked by insects, as is the *R. pseudacacia*. Wood is heavy, hard, close-grained, susceptible of a high polish. Specific gravity, 0.6740 ; ash, 0.80.

XXIV. VITACEÆ. The Vine Family.

123. VITIS LABRUSCA. (Linn.)

Fox Grape.

Common in low, rich grounds.   Stem from fifteen to thirty feet long, straggling over bushes and small trees.   Flowers in June. Fruit in September.

124. VITIS ÆSTIVALIS. (Michaux.)

Little Grape.   Summer Grape.   Common Wild Grape.

There are several varieties of this species, and it is the tallest climber of all our grape-vines, sometimes reaching sixty feet in length.   Flowers in June.   Fruit in October.

124. AMPELOPSIS QUINQUEFOLIA. (Michaux.)

Virginia Creeper.   American Ivy.

Quite common ; growing from ten to fifty feet long ; adhering to trees and walls.   One of the most ornamental of the climbers.

XXV. RHAMNACEÆ. The Buckthorn Family.

125. RHAMNUS CATHARTICUS. (Linn.)

This is introduced, and has become quite naturalized.   It is a low tree, or shrub, with a grayish bark, and is much used for hedges, as it will bear severe pruning.   Flowers in May.   Fruit in October.   The berries are used medicinally as a cathartic.   The *R. ainifolia*, Alder-leaved Buckthorn, is found in moist grounds, and grows from four to six feet high.   Flowers without petals, and fruit black.

XXVI. ACERACEÆ. The Maple Family.

126. ACER RUBRUM. (Linn.)

Red Maple.   Scarlet Maple.   Swamp Maple   Soft Maple.   Water Maple.

Grows from forty to sixty feet high, and from one to two feet or more in diameter.   Wood white or slightly tinted with red, hard, close-grained, easily worked: a little heavier than White Maple, and not as valuable as the Sugar Maple as fuel.   Largely used in cabinet-making, gun-stocks, wooden ware.   This species furnishes the Curled and Bird's-eye Maple for cabinet work.   Specific gravity, 0.6178; ash, 0.37.

127. ACER DASYCARPUM. (Ehrh.)

White Maple.   Silver-leaved Maple.   Soft Maple.

Grows from thirty to sixty feet high, and two feet or more in diameter.   This is one of the most rapid growing of all our maples.   Wood white, fine-grained, hard, susceptible to polish.   Used for cheap furniture, flooring, etc.   Specific gravity, 0.5269 ; ash, 0.33.

128. ACER SACCHARINUM. (WANGENHEIM.)

Sugar Maple. Sugar Tree. Hard Maple. Rock Maple.

Grows from fifty to eighty feet or more in height, and two to three feet in diameter. One of our most rapid-growing and valuable trees. Makes an excellent fuel. Wood heavy, hard, strong, tough, close grained, susceptible of a good polish. Used in the manufacture of furniture, shoe lasts and pegs, saddletrees, for interior finish and flooring. Maple sugar is principally made from this species. The ashes of the wood, rich in alkali, yield large quantities of potash. Specific gravity, 0.6912; ash, 0.36.

129. ACER PENNSYLVANICUM. (LINN.)

Striped Maple. Moose Wood. Striped Dogwood. Goose-foot Maple.

A small tree, with light green bark, striped with darker lines. Wood light, soft, close-grained, satiny. Sometimes cultivated as an ornamental tree. Specific gravity, 0.5299; ash, 0.36.

130. ACER SPICATUM. (LAMARCK.)

Mountain Maple.

Grows from six to ten feet high, but more often a shrub. Grows in rocky places. Color of wood, light brown tinged with red. Wood light, soft, close grained. Specific gravity, 0.5330; ash, 0.43.

XXVII. ANACARDIACEÆ. THE SUMAC FAMILY.

131. RHUS TYPHINA. (LINN.)

Staghorn Sumach.

A common, low shrub, but sometimes a tree from ten to twenty feet high. This is the largest and handsomest species of this genus. Wood orange-colored, brittle, soft, satiny, susceptible of a good polish. Used for inlaying cabinet work. Bark and leaves astringent, rich in tannin. Used medicinally and as a dye, and in dressing skins. Flowers in June. Fruit in September and October. Specific gravity, 0.4357; ash, 0.50.

132. RHUS GLABRA. (Linn.)

Common or Smooth Sumach.

A shrub, growing from three to eight feet high. Fruit red, but in an open and spreading cluster. Flowers in June. Fruit in September and October.

133. RHUS COPALLINA. (Linn.)

Dwarf Sumach.

A shrub, from eight to ten feet in height, found on the borders of woods. Wood soft, coarse-grained, satiny, susceptible of a good polish. Leaves and bark astringent, rich in tannin. Fruit acid and astringent, and used medicinally in the form of decoctions, fluid extracts, etc. Specific gravity, 0.5273; ash, 0.60.

134. RHUS VENENATA. (De Candolle.)

Poison Sumach. Poison Elder. Swamp Dogwood.

A rather handsome shrub, sometimes a small tree from eight to ten feet high. A worthless shrub, and exceedingly poisonous to many persons, but some can handle it with impunity. Opposite effects are sometimes produced on different members of the same family. The poisonous effect is owing to the presence of a volatile principle, *Toxicodendric acid.* Wood soft, coarse-grained. The white, milky sap turns black in drying, and yields a lacquer similar to the Japan varnish, which is supposed to be obtained from a similar species. Specific gravity, 0.4382; ash, 0.64.

135. RHUS TOXICODENDRON. (Linn.)

Poison Ivy. Poison Oak. Poison Vine.

Usually a clinging plant, but sometimes with stem two to five or six feet high. Less poisonous than the *R. venenata.* Always readily distinguished by its leaves and fruit. Flowers in May and June. Fruit in September.

XXIX. RUTACEÆ. The Rue Family.

136. XANTHOXYLUM AMERICANUM. (Miller.)

Prickly Ash. Yellow Wood. Toothache Tree.

Shrub from four to six feet high, but when cultivated and trimmed forms a tree sometimes twenty feet high. Wood light, coarse-grained, soft. The bark is an active stimulant, and is a popular remedy for toothache. All parts of the plant are aromatic.

137. AILANTHUS GLANDULOSA. (Desf.)

Tree of Heaven. Tallow Tree. Chinese Sumach.

Not a native. Cultivated as a shade tree. Grows from thirty to sixty feet high. Native of China. Flowers in June. Fruit in September and October. Of rapid growth. Wood brittle, not very hard, and takes a good polish.

XXX. TILIACEÆ. The Linden Family.

138. TILIA AMERICANA. (Linn.)

Basswood. Lime Tree. American Linden. Beech Tree. Whitewood.

Grows from forty to eighty feet high, and from two to three feet in diameter. Wood light, soft, not strong, close-grained, easily worked. Used in the manufacture of woodenware, paper pulp, interior decoration, etc. The inner bark consists of long, tough fibres, easily separated, and used in mats, cordage, etc. Flowers rich in honey. Specific gravity, 0.4525 ; ash, 0.55.

XXXI. BERBERIDACEÆ. The Barberry Family.

139. BERBERIS VULGARIS. (Linn.)

Barberry.

Shrub, from three to ten feet high, producing numerous suckers. Native of Europe, but thoroughly naturalized here. The berries and inner bark are used medicinally. Flowers in May. Fruit in October.

XXXII. MAGNOLIACEÆ. THE MAGNOLIA FAMILY.

140. MAGNOLIA GLAUCA. (LINN.)

Laurel. Small Magnolia. Sweet Bay. Swamp Laurel. Beaver Tree. White Bay.

Shrub or small tree, from four to twenty feet high. One of our most beautiful ornamental shrubs. Wood light, soft, close-grained, compact. Bark and root used medicinally. Specific gravity, 0.5035 ; ash, 0.47.

141. LIRIODENDRON TULIPIFERA.

Tulip Tree. White Wood. Canoe Wood. Yellow Poplar.

One of our largest and most valuable trees, growing from eighty to one hundred and twenty feet high, and from two or three to five or six feet in diameter. Wood light, not strong, very close straight-grained, compact, easily worked, and is one of the few kinds of wood that will shrink endways of the grain when seasoning. Used in interior finish, carriages, shingles, etc. *Liriodendrin*, a stimulant tonic with diaphoretic properties, is obtained by macerating the inner bark, especially of the root. Specific gravity, 0.4230 ; ash, 0.23.

ADDENDA.

Since this article was written my attention has been called to the following additions :

ORDER SALICACEÆ.

POPULUS DILITATA. (AIT.]

Lombardy Poplar. Italian Poplar.

Formerly a favorite ornamental tree, but not now common.

ORDER URTICACEÆ.

MACLURA AURANTIACA. (NUT.)

Osage Orange. Bow-wood. Bodock.

Grows from fifteen to twenty feet high. Wood very hard and durable. Used by the aborigines for making bows. Was called by the early French settlers Bois d'Arc, from whence Bodock.